Otto Moeschlin
Eugen Grycko

Experimental Stochastics in Physics

W0091526

Otto Moeschlin   Eugen Grycko

Experimental Stochastics in Physics

With 26 Figures, a CD-ROM
and an Installation Manual

 Springer

Otto Moeschlin
Eugen Grycko

FernUniversität Hagen
Fachbereich Mathematik
Lützowstr. 125
58084 Hagen, Germany
e-mail: otto.moeschlin@fernuni-hagen.de

Library of Congress Control Number: 2005936832

Mathematics Subject Classification (2000):
11K45, 60Fxx, 62Fxx, 62Gxx, 62Jxx, 82Bxx, 82Dxx

ISBN-10 3-540-28362-5   Springer Berlin Heidelberg New York
ISBN-13 978-3-540-28362-1   Springer Berlin Heidelberg New York

Springer is a part of Springer Science+Business Media

springer.com

© Springer-Verlag Berlin Heidelberg 2006
Printed in Germany

Typesetting: Authors' input files edited and reformatted by LE-TEX Jelonek, Schmidt & Vöckler
GbR, Leipzig
Production: LE-TEX Jelonek, Schmidt & Vöckler GbR, Leipzig
Cover design: KünkelLopka, Heidelberg

Printed on acid-free paper   46/3142YL - 5 4 3 2 1 0

Preface

The electronic monograph 'Experimental Stochastics in Physics' might be seen as a continuation of the electronic monograph 'Experimental Stochastics', which first appeared in 1998 with a second edition in 2003. Thereby 'Experimental Stochastics' has to be understood as a synonym for 'Stochastic Simulation'.

The intention is and was to get computer-experimental insights into virtual molecular systems, that is, to estimate macro quantities of such systems in the statistical word sense based on stochastic dynamical models established on the computer.

Although mathematics is strongly used, the working method is not mathematical, rather the one of natural sciences; that is, the results are gained through experimentation. The physical experiment is replaced by the stochastic one, and the measuring devices by stochastical/statistical utensils. Indeed, 'Experimental Stochastics in Physics' makes use of statistical methods which are based implicitly or explicitly on the strong law of large numbers.

In this sense it is natural to focus especially those parts of physics on which are related to stochastics. This indeed characterizes the contents of 'Experimental Stochastics in Physics'.

'Experimental Stochastics in Physics' allows insights which cannot be established by laboratory physics. This holds true especially for computer-experiments from non-real physics; by this we understand experiments which are not covered by natural laws.

In the present monograph the reader is not introduced to the realm of random generators or on how to generate sample realizations which follow a certain probabilty distribution; for such concepts we refer to 'Experimental Stochastics'.

The monograph 'Experimental Stochastics in Physics' comes on a CD-Rom with a number of stochastic-physical experiments especially from the field of 'Statistical Mechanics', together - for better understanding - with the text from the accompanying brochure.

'Experimental Stochastics in Physics' is intended for all persons interested in physics or stochastics/statistics. From the methodology applied 'Experimental Stochastics in Physics' is well suited for mathematicians, statisticians, scientists, not only physicists, as well as engineers and representatives from the field of Operational Research with an interest in (stochastic) simulation. The experiments offered allow easy access, and so 'Experimental Stochastics in Physics' may also be used successfully for teaching purposes.

'Experimental Stochastics in Physics' has been developed together with students and former students enrolled at the Department of Probability Theory and Statistics of the University of Hagen. Our thanks go to all who have contributed to the successful outcome of 'Experimental Stochastics in Physics'. We are particularly grateful to Professor Michael Eastham of Cardiff University for his help with the translation into English. We also thank Frau Gisela Soentgen who prepared the TeX manuscript. For a complete list we refer to 'About', which can be displayed on screen through the Top(-Shell).

Hagen, Autumn 2005   *Otto Moeschlin, Eugen Grycko*

The portraits decorating the Top(-Shell) are those of James C. Maxwell and Ludwig Boltzmann.

Contents

Introduction

The (electronic) monograph 'Experimental Stochastics in Physics' deals with random experiments on the computer from the field of molecular dynamics.

Based on stochastic dynamic models of molecular movement established on the computer, macro-quantities of interest are estimated in the statistical word sense.

Since the computer processes a great amount of data, the applied statistical methods are especially asymptotic ones; implicitly or explicitly the statistical argumentation is based on laws of large numbers.

Decisive for the physical model of molecular movements is the development of the dynamics, which is addressed several times within 'Experimental Stochastics in Physics', while the generation of random data according to predetermined probability distributions is not the subject of the present monograph. The reader is referred to 'Experimental Stochastics', cf. Moeschlin et al. (2003).

The molecular system on which our investigations and also our generalizations are based is the Boltzmann system of moving molecules on which Newtonian dynamics is imposed. Note that the concept of energy conservation is never given up, the one of momentum conservation only in Chapter 12.

A main result of Statistical Mechanics is certainly the so-called Maxwell Hypothesis which states that the velocities and momenta, respectively, of the molecules of a fluid (gas) are distributed according to a normal distribution with the variance being determined by the temperature. While the distributional aspect of Maxwell's Hypothesis might be seen as a statistical/stochastical (i. e. a mathematical) statement, the relation between the variance and the temperature is an intrinsic physical insight. From a mathematical point of view the latter might be reduced to the fact that temperature has to be a scalar quantity.

A direct way to prove the Maxwell Hypothesis by laboratory experiments does not exist, not even today. In contrast, the computer experimentation allows us to examine the Maxwell Hypothesis directly; to this end the so-called Boltzmann system of moving molecules subject to Newtonian dynamics is developed for implementation on the computer, the latter being understood as a virtual laboratory; the measuring devices are the various estimators.

Indeed, the experiments show that the estimated velocity and momentum distributions of a molecule are ergodic, i. e. they become stationary, and are centered, rotationally symmetric normal distributions according to Maxwell's Hypothesis. The estimated variance of such a stationary distribution yields a temperature of the virtual system of molecules installed on the computer.

Of course, any statistical aspect of a Boltzmann system of moving molecules may be investigated by statistical methods based on data from computer experimentation, for instance the counting process, induced by the collision process of molecules, reveals itself as Poissonian. Indeed, there are arguments, suggesting the Poissonity of the counting process.

But indeed a proof should – if the possibility exists – be given experimentally or alternatively computer-experimentally. Computer experimentation also permits the estimation of pressure, which opens up a possibility of examining the equation of state interrelating pressure and temperature. We refer to Chapters 3–5.

The computer as a virtual laboratory allows more than only resetting a dynamics from real physics; in our virtual laboratory, experiments are possible which have no real counterpart; that is, experiments can be carried out which contradict the laws of nature, cf. Chapter 6. One has to take it that (only) experiments of this kind permit the study of interdependences, that is, the interrelationship of the various concepts and laws of nature.

In any case the discovery of the Maxwell Hypothesis is a great physical perception; however, the question rises whether the Maxwell Hypothesis is a window to a more comprehensive natural law of the same kind. A way to such an analysis is offered by Theorem 2.2.3, cf. Mackey (1992), yielding the optimizer of the Boltzmann–Gibbs entropy, which permits us to reformulate the Maxwell Hypothesis in a generalized form, called here the Entropy Principle. The possibility of extending the concept of the Maxwell Hypothesis is based on the fact that the formally introduced measurable, non-negative function H in Theorem 2.2.3 may be interpreted as Hamiltonian.

The formalism needed to realize this extended concept of the Maxwell Hypothesis is based on a family of momentum distributions parametrized by $\beta, 0 < \beta \in \mathbb{R}$, called here the family of entropic distributions. The accurate specification of β yields – according to the Maxwell Hypothesis – the accurate temperature.

Generalizations of the Newtonian concepts of momentum and energy which – apart from the relativistic dynamics and possibly the one in connection with discrete momentum and energy in Chapter 11 – are not realized in nature, may indeed yield more sophisticated insights.

It is a well-known experience from mathematics that a generalization to higher dimensions enables us to differentiate between phenomena which seem at first to coincide, for instance straight lines being skew should not be investigated in linear spaces with dimension less than 3.

The decisive questions – not answered by the Maxwell Hypothesis – are: For which types of modified Hamiltonians and which types of modified momenta as functions of the velocity, determining the behavior of a system of moving molecules, is the generalized Maxwell Hypothesis (i. e. the Entropy Principle) fulfilled?

In other words, is the validity of the Maxwell Hypothesis only given for the special case of momentum and energy as defined by Newton? To answer this question empirically in Chapter 7 (M, M)-dynamics is first introduced, which means a natural generalization of Newtonian dynamics. Are there special relations between kinetic energy and momentum to be met such that the collision process of the molecules is not impeded?

Does the validity of the Entropy Principle depend on a special organizational form of the energy splitting? Does the energy splitting have to be caused by an impact of two molecules? Can this causality be given up? What role does the momentum exchange direction play?

Empirical answers are given in Chapters 8 and 9, where non-separating (M, B)-dynamics as well as an acausal variant of (M, M)-dynamics are examined, respectively.

Of course, there is first the distributional aspect, that is, the question whether the Entropy Principle remains valid, but also the aspect whether the equation of state interrelating temperature and pressure still holds true.

Chapters 10 and 11 treat relativistic dynamics, and a dynamics where energy is defined as the sum of the absolute values of the momentum components, respectively, i. e. the functional forms of energy and momentum are enlarged. In the latter we have in addition an alteration with respect to the topology of the momentum space, which is assumed to be discrete, with the consequence that momentum vectors are restricted to a lattice, which leads to a stochastical concept determining the momentum exchange directions. Moreover, the causality of the momentum exchange is given up as in Chapter 9. A natural question in the context of Chapter 11 is whether this model can be understood as a 'mechanical' model establishing the theoretically postulated equilibrium probabilities of the excited energy levels of a harmonic oscillator from quantum mechanics?

After having altered the definition of momentum as well as the organisational form of momentum exchange several times in Chapters 7–11 we finally ask whether it is possible to give up the concept of momentum exchange; i. e. can this concept be substituted by a stochastical rule determining the energy splitting? Answers are given in Chapter 12.

In any of Chapters 7–12 we offer (at least) four typical experiments. With the first experiment within a chapter we investigate the existence and type of the ergodic distribution. The second experiment examines the existence of a temperature as a scalar quantity based on the comparison of the empirical, i. e. estimated (ergodic), distribution with the theoretically defined family of the entropic distributions. That is, it examines the validity of the Entropy Principle.

The subject of the third experiment within a chapter is the validity of the equation of state, interrelating pressure and temperature, which for reasons of definition requires the invariance of pressure. The latter is indeed not always the case.

Finally, with the fourth experiment of a chapter the equipartition of energy is addressed.

1 Boltzmann System of Moving Molecules

1.1 Contributions of Maxwell and Boltzmann

Statistical Mechanics is associated in particular with the names of two nineteenth century physicists, the Briton James Clerk Maxwell (13 June 1831 – 5 Nov 1879) and the Austrian Ludwig Boltzmann (20 Feb 1844 – 5 Oct 1906).

In 1860 Maxwell formulated the so-called *Maxwell Hypothesis* which states in a slightly generalized form that the velocity vectors of a gas molecule (micro-constituent of a fluid)

(a) follow in an equilibrium state a centered normal distribution:

$$(1.1.1) \qquad N(0, \sigma_v^2) \otimes \ldots \otimes N(0, \sigma_v^2) \; = \; N(0, \sigma_v^2 I_d)$$

on the measurable space $(\mathbb{R}^d, \mathcal{B}^d)$, $\mathcal{B}^d$ being the Borel σ-field over $\mathbb{R}^d$. The left side of (1.1.1) denotes the d-fold product of the normal distribution $N(0, \sigma^2)$ on $(\mathbb{R}, \mathcal{B})$ with mean $0 \in \mathbb{R}$ and variance $\sigma_v^2 \in \mathbb{R}_+$, while $N(0, \sigma_v^2 I_d)$ signifies the d-dimensional normal distribution with mean $0 \in \mathbb{R}^d$ and the covariance matrix $\sigma^2 I_d$ (I_d: d-dimensional identity matrix) for $d = 2, 3$.

(b) the variance σ_v^2 is – up to a multiplicative constant – determined by the temperature T in [K] (Kelvin):

$$(1.1.2) \qquad \sigma_v^2 \; = \; \frac{k_{\mathrm{B}} \cdot T}{m}$$

with m as the mass of a molecule and $k_{\mathrm{B}} = 1.38 \cdot 10^{-23}$ [J/K] as the so-called Boltzmann constant.

The probability measure (1.1.1) with σ_v^2 defined by (1.1.2) is called the **Maxwell–Boltzmann velocity distribution** of a molecule.

The development of the theory of Statistical Mechanics by Ludwig Boltzmann is based on a model of a gas (fluid) in a container represented by N moving molecules as hard disks of mass $m > 0$ subject to Newtonian dynamics which entails that momentum and mechanical energy remain constant over time. We are speaking here of the *Boltzmann system of moving molecules*.

The impacts between molecules or between a molecule and the boundary of the container are elastic, the total kinetic energy at any collision being totally converted into kinetic translation energy, i. e. no conversion into rotational energy takes place.

Statement (b) of the Maxwell Hypothesis, cf. (1.1.2), is an intrinsic physical insight which cannot be tested or reproduced on the computer, while statement (a) should – as a consequence of the prevailing Newtonian dynamics – remain valid even if physical reality is substituted by a corresponding computer reality which will be our research here.

To this end we develop a stochastic dynamic model on the computer 'imitating' the Boltzmann model of moving molecules (for details see Sections 1.6–1.8) and check in experiment E 1.1 whether an arbitrary initial velocity distribution of a molecule tends – as a consequence of the impacts of the molecules – towards a *stationary* distribution, being the one postulated by Maxwell, cf. (1.1.1). We also speak of this stationary distribution as the *ergodic* or the *equilibrium* velocity distribution, cf. Section 2.2.

Experiment E 1.1 confirms statement (a) of the Maxwell Hypothesis.

This gives rise to addressing the same problem context in Chapters 7–12; in Chapter 10 for instance with respect to relativistic dynamics, while the dynamics in Chapters 7–9 have no real counterpart.

Typical for Statistical Mechanics is that the total kinetic energy is represented as an expectation with respect to the velocity distribution and to the distribution of momentum, respectively.

1.2 Expected Kinetic Energy

Let a gas of N molecules with mass $m > 0$ confined to a container B be given as a Boltzmann system of moving molecules, cf. Section 1.1, the gas being considered either in the 3-dimensional real world or in a 2-dimensional virtual laboratory on a computer. This prompts us to introduce notions from the theory of moving molecules in $\mathbb{R}^d$, $d = 2, 3$. In particular, the container B is a set $B \subset \mathbb{R}^d$ having a non-void topological interior.

If the velocity vectors

$$(1.2.1) \qquad v^{(j)} = (v_1^{(j)}, \ldots, v_d^{(j)}) \in \mathbb{R}^d, \qquad j \in \mathbb{N}_N,$$

are known, the average kinetic energy E_{av}^N of N molecules is given by

$$(1.2.2) \qquad E_{\mathrm{av}}^N := \frac{1}{N} \sum_{j=1}^{N} \frac{m}{2} \langle v^{(j)}, v^{(j)} \rangle = \frac{1}{N} \sum_{j=1}^{N} \frac{m}{2} (v_1^{(j)^2} + \ldots + v_d^{(j)^2}),$$

$\langle ., . \rangle$ denoting the standard scalar product on $\mathbb{R}^d$.

It is the advantage of the Boltzmann system of moving molecules being imitated on the computer that these velocities are known, being generated artificially according to a given initial distribution and then computed again at every collision, step by step.

Of course, the velocities of the molecules are not known in reality, and that is why in Statistical Mechanics energy will be represented as an expectation with respect to the velocity distribution or to the distribution of momentum. To this end we introduce the velocity space of the j-th molecule as

$$(1.2.3) \qquad (\mathbb{V}^{(j)}, \mathcal{V}^{(j)}, P^{(j)}) \; = \; (\mathbb{V}, \mathcal{V}, P) \; = \; (\mathbb{R}^d, \mathcal{B}^d, P), \qquad j \in \mathbb{N},$$

$\mathcal{B}^d$ being the Borel σ-field over $\mathbb{R}^d$.

We now idealize the Boltzmann system of moving molecules insofar as we accept countably many molecules. Let $(\Omega, \mathcal{A}, Q)$ be a probability space with

$$\Omega := \underset{j=1}{\overset{\infty}{\times}} \mathbb{V}^{(j)}$$

$$\mathcal{A} = \bigotimes_{j=1}^{\infty} \mathcal{V}^{(j)}$$

$$Q = \bigotimes_{j=1}^{\infty} P^{(j)}$$

and $\mathbb{V}^{(j)} = \mathbb{V}$, $\mathcal{V}^{(j)} = \mathcal{V}$, $P^{(j)} = P$, $j \in \mathbb{N}$.

The velocity vector

$$V^{(j)} := (V_1^{(j)}, \dots, V_d^{(j)})$$

of the j-th molecule is the projection of the probability space $(\Omega, \mathcal{A}, Q)$ onto $\mathbb{V}^{(j)} = \mathbb{V}$.

Ω is the set of all sequences $\omega := (v_j)_{j \in \mathbb{N}}$, $v_j \in \mathbb{R}^d$. Therefore we have

$$(1.2.4) \qquad V^{(j)}(\omega) \; = \; v_j, \qquad (v_j)_{j \in \mathbb{N}} \; = \; \omega \in \Omega.$$

Structuring the velocity space by

$$(1.2.5) \qquad (\mathbb{V}, \mathcal{V}, P) := \left(\underset{i=1}{\overset{d}{\times}} \mathbb{V}_i, \bigotimes_{i=1}^{d} \mathcal{V}_i, P \right),$$

the components $V_i^{(j)}$ of $V^{(j)}$ are given by

$$(1.2.6) \qquad V_i^{(j)} \; = \; \pi_i \circ V^{(j)}, \qquad i = 1, \dots, d,$$

where π_i, denotes the projection from $\mathbb{V}$ onto $\mathbb{V}_i$, $i = 1, \dots, d$.

Accordingly, the components $V_i^{(j)}$ are distributed according to P_{π_i}, i.e. the marginal distributions of P with respect to i, $i = 1, \dots, d$.

Therefore the expected kinetic energy of a molecule is given by

$$(1.2.7) \qquad E_{\mathrm{e}} \; = \; \sum_{i=1}^{d} \frac{m}{2} \int (\mathrm{id}_{\mathbb{V}_i})^2 \, \mathrm{d} P_{\pi_i}$$

or in a more suggestive form by

$$(1.2.8) \qquad E_e \;=\; \sum_{i=1}^{d} \frac{m}{2} \int v_i^2 \; dP_{\pi_i} \,.$$

If the probability measure P is a product, i. e. if

$$P \;=\; \bigotimes_{i=1}^{d} P_0 \,,$$

with P_0 being the ergodic normal distribution $N(0, \sigma_v^2)$ with variance σ_v^2 postulated by Maxwell, we get for the expected kinetic energy of a molecule

$$(1.2.9) \qquad E_e \;=\; d\frac{m}{2}\sigma_v^2, \qquad d = 2,3 \,.$$

1.3 Averaging the Kinetic Energies

By the model of random vectors introduced in Section 1.2, the random vectors $V^{(j)}$, $j \in \mathbb{N}$, are identically distributed and stochastically independent. We moreover assume that these random vectors are quadratically integrable, which is a standard model assumption.

Therefore the random vectors $V^{(j)} = (V_i^{(j)})_{i \in \mathbb{N}_d}$ satisfy the strong law of large numbers:

$$(1.3.1) \qquad \lim_{N \to \infty} \frac{1}{N} \sum_{j=1}^{N} (V_i^{(j)})^2 \;=\; \int v_i^2 \; dP_{\pi_i} \,, \quad i \in \mathbb{N}_d \,, \qquad Q\text{-a.e.}$$

which entails

$$(1.3.2) \qquad \lim_{N \to \infty} E_{av}^N \;=\; \lim_{N \to \infty} \frac{1}{N} \sum_{i=1}^{d} \sum_{j=1}^{N} \frac{m}{2}(V_i^{(j)})^2$$

$$=\; \frac{m}{2} \sum_{i=1}^{d} \int v_i^2 \; dP_{\pi_i} \;=\; E_e \qquad Q\text{-a.e.}$$

1.4 Temperature and Energy

Formula (1.2.7) represents the expected kinetic energy of a molecule for the stationary velocity distribution

$$\bigotimes_{i=1}^{d} N(0, \sigma_v^2) \,.$$

By the statement (b) of the Maxwell Hypothesis, cf. Section 1.1, and Formula (1.2.9) we obtain

$$(1.4.1) \qquad E_e = \frac{d}{2}m\sigma_v^2 = \frac{d}{2}\,k_{\mathrm{B}}\cdot T\,,$$

which expresses a relation between temperature and energy, going back to Ludwig Boltzmann. This result is meaningful insofar that, in Newtonian mechanics, the notion of temperature does not exist.

Solving (1.4.1) with respect to T we get using (1.3.2)

$$(1.4.2) \qquad T = \frac{2}{d\cdot k_{\mathrm{B}}}\,E_e = \frac{2}{d\cdot k_{\mathrm{B}}}\,\lim_{N\to\infty}E_{\mathrm{av}}^N \qquad Q\text{-a.e.}$$

This shows that

$$(1.4.3) \qquad \hat{T}_N := \frac{2}{d\cdot k_{\mathrm{B}}}\,E_{\mathrm{av}}^N = \frac{1}{d\cdot k_{\mathrm{B}}}\,\frac{1}{N}\sum_{i=1}^{N}m\,\langle v_i,v_i\rangle$$

is a strongly consistent estimator for the temperature.

Remark 1.4.1
If the ergodic distribution is already attained, Formula (1.4.3) allows us to compute reliable estimates for the temperature. Obviously, statement (b) of the Maxwell Hypothesis is involved in the deduction of (1.4.3).

The inverse problem to generate velocities of the molecules according to the ergodic velocity distribution can be solved by, for example by the Box–Müller procedure, cf. Moeschlin, Grycko, Pohl, Steinert (2003), which delivers realizations of the normal distribution using random generators. To adjust a certain temperature, the variance σ^2 in (1.1.1) is calculated according to statement (b) of the Maxwell Hypothesis.

1.5 Density of the Momentum Distribution

We consider here again the Boltzmann system with N moving molecules. In 1.2 the velocity space of a molecule was introduced as a probability space. In Statistical Mechanics the distribution of the momentum plays an even more important role than the distribution of the velocity. We therefore introduce the momentum space of the j-th molecule as

$$(1.5.1) \qquad \begin{aligned}(\mathbb{U}^{(j)},\mathcal{U}^{(j)},P_U) &:= \left(\bigtimes_{i=1}^{d}\mathbb{U}_i^{(j)},\bigotimes_{i=1}^{d}\mathcal{U}_i^{(j)},P_U\right) = \left(\mathbb{R}^d,\mathcal{B}^d,P_U\right) \\ &=: \left(\mathbb{U},\mathcal{U},P_U\right),\quad j\in\mathbb{N},\end{aligned}$$

$\mathcal{B}^d$ being the d-dimensional Borel σ-field.

There, P_U represents the distribution of the identically distributed random vectors of momentum $U^{(j)} = (U_1^{(j)}, \ldots, U_d^{(j)})$, $j \in \mathbb{N}$, with d being either 2 or 3.

The elements of $\mathbb{U}^{(j)}$ are represented by $u^{(j)} := (u_1^{(j)}, \ldots, u_d^{(j)})$, i.e. by the possible vectors of momentum of the j-th molecule.

We dispense with presenting the further developments which completely correspond those given in Section 1.2 for the velocity case.

Instead of the distribution (1.1.1) we now have

$$(1.5.2) \qquad \bigotimes_{i=1}^{d} N(0, \sigma_u^2) = N(0, \sigma_u^2 I_d)$$

where σ_u^2 denotes the variance of the distribution of a component of a random momentum vector of a molecule; the variance of the distribution of a component of a random velocity vector is denoted by σ_v^2.

The density of the equilibrium distribution $N(0, \sigma_v^2)$ of a component $V_i^{(j)}$, $i \in \mathbb{N}_d$, of a random velocity vector of a molecule is given by

$$(1.5.3) \qquad f(v_i) = (2\pi\sigma_v^2)^{-\frac{1}{2}} \exp(-v_i^2 / 2\sigma_v^2), \qquad v_i \in \mathbb{R}.$$

Based on the relation

$$U_i^{(j)} = m V_i^{(j)}, \quad i \in \mathbb{N}_d, \ j \in \mathbb{N},$$

between the components of random velocity and momentum vectors, the density g of the equilibrium distribution of a component $U_i^{(j)}$ of the random momentum vector $U^{(j)}$, $i \in \mathbb{N}_d$, of a molecule can be determined by the rule of density transformation as

$$(1.5.4) \qquad g(u_i) = m^{-1}(2\pi\sigma_v^2)^{-\frac{1}{2}} \exp(-u_i^2 / 2m^2\sigma_v^2)$$

$$= (2\pi\sigma_u^2)^{-\frac{1}{2}} \exp(-u_i^2 / 2\sigma_u^2), \qquad u_i \in \mathbb{R},$$

with

$$(1.5.5) \qquad \sigma_u^2 = m^2\sigma_v^2 = m^2 k_{\mathrm{B}} T / m = m \cdot k_{\mathrm{B}} \cdot T.$$

Therefore the distribution of a component of the random momentum vector of a molecule is given by

$$(1.5.6) \qquad N(0, \sigma_u^2) = N(0, m \cdot k_{\mathrm{B}} \cdot T).$$

Now (1.5.6) gives rise to a reformulation of the Maxwell Hypothesis **for the momenta** instead of for velocities as in (1.1.1) and (1.1.2).

Principle 1.5.1 (Maxwell Hypothesis for momenta)

The ergodic momentum distribution of a Boltzmann system of molecules from $\mathbb{R}^d$, $d = 2, 3$, has the following two properties.

(a) It is a centered normal distribution

$$\text{(1.5.7)} \qquad \left(N(0, \sigma_u^2)\right)^d \;=\; N(0, \sigma_u^2 I_d)$$

on the measurable space $(\mathbb{R}^d, \mathcal{B}^d)$.

(b) The variance σ_u^2 is – up to a multiplicative constant – determined by the temperature in $[K]$ (Kelvin):

$$\text{(1.5.8)} \qquad \sigma_u^2 \;=\; k_{\mathrm{B}} \cdot T \cdot m \,.$$

The probability measure (1.5.7) with σ_u^2 defined by (1.5.8) is called the **Maxwell–Boltzmann momentum distribution** of a molecule.

Instead of the definition of the average kinetic energy E_{av}^N given in (1.2.2) based on velocity vectors, this definition can also be given in terms of momentum vectors:

$$\text{(1.5.9)} \qquad E_{\mathrm{av}}^N \;=\; \frac{1}{N} \sum_{j=1}^{N} \frac{1}{2m} \, \langle u^{(j)}, u^{(j)} \rangle \,.$$

The same holds true for the expected kinetic energy E_{e} of a molecule for which, by (1.2.9), we now obtain

$$\text{(1.5.10)} \qquad E_{\mathrm{e}} \;=\; d \frac{1}{2m} \sigma_u^2 \,.$$

(1.3.2) remains true independently of how the definitions of E_{av}^N and E_{e}, respectively, are given:

$$\text{(1.5.11)} \qquad \lim_{N \to \infty} E_{\mathrm{av}}^N \;=\; E_{\mathrm{e}} \qquad Q\text{-a.e.}$$

1.6 Design of the Experiment

With a 2-dimensional, virtual experiment on the computer, imitating the Boltzmann system of moving molecules, the distributional aspect of the Maxwell Hypothesis, statement (a), cf. (1.1.1), will be examined in the sense of a temporal evolution of a given initial distribution of the velocities towards the normal distribution

$$\text{(1.6.1)} \qquad N(0, \sigma_v^2) \otimes N(0, \sigma_v^2) \;=\; N(0, \sigma_v^2 I_2)$$

postulated by Maxwell as the ergodic distribution of the random velocity vectors.

We consider a 2-dimensional circular container B with radius $R > 0$, i.e.

$$B \;:=\; \left\{ (x_1, x_2) \in \mathbb{R}^2 \,\middle|\, x_1^2 + x_2^2 \leq R \right\} \,.$$

At the outset N non-overlapping molecules of mass $m > 0$, considered as hard discs with radius $r > 0$, are uniformly distributed in B, while the initial distribution of the velocity vectors can be selected by the experimenter.

The kernel of the experiment is – from a technical viewpoint – the concept of the implemented dynamics of the molecules interacting both between themselves and with the boundary of the container, cf. 1.8 'Kinetic Dynamics'!

In order to show that the ergodic velocity distribution on $\mathcal{V} = \mathcal{B}^2$ is of the type (1.6.1), we estimate the densities of the projections of the velocity distribution onto the linear subspaces L_φ of $\mathbb{V}$, determined by the polar angle φ, $0° \leq \varphi \leq 360°$, with the result that these projections are statistically identical, equal to $N(0, \sigma_v^2)$.

In experiment E 1.1 we estimate the densities of the projections of the velocity distribution onto a variable linear subspace L_φ of $\mathbb{V}$, being determined by a polar angle φ which can be altered by the experimenter.

In the same way the momentum distribution is also examined in experiment E 1.1.

1.7 Estimation Methods

We discuss here the estimation of the velocity distribution; the estimation of the momentum distribution is established similarly.

The densities of the projections of the unknown velocity distribution on $\mathcal{V} = \mathcal{B}^2$ onto the linear subspaces L_φ of $\mathbb{V}$ are estimated by means of a kernel density estimator based on the velocity data of the moving molecules.

The kernel density estimator is a tool from non-parametric statistics which only yields the numerical values of the density (that is, it does not specify a parameter within a parametrized class of distributions), such that the graph of the density can be displayed . The comparison, i. e. the coincidence with the density of the normal distribution $N(0, \sigma_v^2)$, where the variance is estimated based on the same velocity data, allows not only to specify the density by a formula but also the actual parameter.

With arguments as used to prove (1.3.1) and (1.3.2), it can be shown that the sequence $(S_{\varphi N}^2)$ of estimators

$$(1.7.1) \qquad S_{\varphi N}^2 := \frac{1}{N} \sum_{j=1}^{N} V_\varphi^{(j)2} \,,$$

where $V_\varphi^{(j)} := \pi_\varphi \circ V^{(j)}$ and π_φ denotes the projection from V onto L_φ, is strongly consistent for the estimation of the second moment of any distribution, and therefore for the variance of the centered normal distribution.

Taken as an estimator based on a finite number of sample values, the estimator (1.7.1) is of minimal variance with respect to the class of centered normal distributions, cf. Moeschlin, Eberl (1982), example 3.6.12.1.

The probability measure (1.6.1) – being a product measure – proves by its rotational symmetry to be the product of its projections onto any pair of orthogonal linear subspaces of $\mathbb{V} = \mathbb{R}^2$.

Indeed, our estimations show that these projections are statistically identical, equal to $N(0, \sigma_v^2)$, i. e. the estimates of the variances differ only slightly. This confirms the Maxwell Hypothesis, statement (a), cf. (1.6.1).

The estimation of the variance σ^2 is only possible when the velocity distribution has become stationary.

On principle, any estimation has to be established for a stationary distribution; nevertheless we estimate the densities during the temporal evolution of the velocity distribution by means of a kernel density estimator.

Indeed, the quality of such estimations depends on the one hand on the velocity with which the velocity distribution is altering, and on the other hand on the amount of data which is required for a satisfactory estimation.

Our experience is that the kernel density estimator yields instructive images of the densities of the velocity distribution, even if they alter.

1.8 Kinetic Dynamics

In this section we describe the dynamics to be implemented on the computer, which is based on the Newtonian axioms. The outcome of a collision between molecules is determined by the laws of conservation of momentum and energy. As within the experiments the molecules are represented by hard discs, it is sensible to restrict ourselves exclusively to translation movements, that is, there is no transformation of kinetic translation energy to rotational energy. In the following we consider a 2-dimensional container B, which is modeled as a not necessarily convex subset of the Euclidean plane: $B \subset \mathbb{R}^2$ with a non-void topological interior. Furthermore we assume that the boundary ∂B of B is a piecewise differentiable curve such that, with the exception of a subset $D \subset \partial B$ of Lebesgue measure zero there is at every point $x \in \partial B \setminus D$ a uniquely determined direction which is orthogonal to the boundary ∂B.

For later reference we generalize the dynamics in comparison to the assumptions already laid down in 1.2 to the case that there are two kinds of molecules, namely N_1 molecules of kind 1 with radius $r_1 > 0$ and mass $m_1 > 0$, and N_2 molecules of kind 2 with radius $r_2 > 0$ and mass $m_2 > 0$; cf. Section 5.2 'Osmotic Pressure'.

Let

$$(x(0); v(0)) := \left(x^{(1)}(0), \ldots, x^{(N)}(0); v^{(1)}(0), \ldots, v^{(N)}(0) \right) \in B^N \times \mathbb{R}^{2N}$$

be an initial (microscopic) state of the system, where $x^{(i)}(0)$ denotes the position and $v^{(i)}(0)$ the velocity of the i-th molecule at the time instant $t = 0$, $i = 1, \ldots, N$. Thereby molecules with indices $i \in \mathbb{N}_{N_1} =: K_1$ are molecules of the first kind, those with indices $i \in \mathbb{N}_{N_1+N_2} \setminus K_1 =: K_2$ are molecules of the second kind.

The microscopic state of the system at time $t > 0$ is given by

$$\Big(x(t); v(t)\Big) \;=\; \Big(x(0) + t\,v(0); v(0)\Big);$$

if no collisions between the molecules and no reflections of a molecule off the boundary have taken place during the time interval $[0, t]$.

The time t_i, $1 \leq i \leq N$, of a potential reflection of the i-th disc off the boundary can be computed as the smallest positive solution of the equation

$$d\Big(x^{(i)}(0) + tv^{(i)}(0), \partial B\Big) \;=\; d_1,$$

d_1 being either r_1 or r_2, depending on the kind of molecules; d denotes the Euclidean distance on $\mathbb{R}^2$, while the expression $d(y, \partial B)$ is defined by

$$d(y, \partial B) \;:=\; \inf\{d(y, a) | a \in \partial B\}$$

for $y \in B$.

The time t_{ij}, $1 \leq i, j \leq N$, of a potential collision between the molecules i and j is the smallest positive solution of the quadratic equation

(1.8.1) $$d\Big(x^{(i)}(0) + tv^{(i)}(0), x^{(j)}(0) + tv^{(j)}(0)\Big) \;=\; \rho_2,$$

ρ_2 being either $2r_1$ or $2r_2$ or $r_1 + r_2$ depending on whether $i, j \in K_1$ or $i, j \in K_2$ or $(i, j) \in K_1 \times K_2 \cup K_2 \times K_1$.

If (1.8.1) has no real or no positive solutions, then the molecules i and j will not collide with each other before the next collision of two other molecules or reflection of a molecule off the boundary has occurred. The time $\bar{t}$ of the next collision or reflection in the system is defined as

(1.8.2) $$\bar{t} \;:=\; \min\Big(\{t_i | i \in \mathbb{N}_N\} \cup \{t_{ij} | i, j \in \mathbb{N}_N, i < j\}\Big).$$

If the i-th molecule reflects off the boundary of the container (such a reflection is interpreted as an instantaneous action, i. e. an action whose time duration has Lebesgue measure 0) at time $\bar{t}$, then the velocity vector $v^{(i)}(\bar{t})$ is reversed at the tangent to ∂B at the point of reflection.

If, however, a collision between the i-th and the j-th molecules occurs at time $\bar{t}$, then the direction of the momentum exchange is given by

(1.8.3) $$e \;:=\; \frac{x^{(i)}(\bar{t}) - x^{(j)}(\bar{t})}{\big|\,x^{(i)}(\bar{t}) - x^{(j)}(\bar{t})\,\big|}.$$

Let u_i, u_j and $\bar{u}_i, \bar{u}_j$ denote the momentum of the molecules i and j immediately before and immediately after the impact at time $\bar{t}$, respectively. Then the conservation of momentum implies that

$$(1.8.4) \qquad \begin{aligned} \bar{u}^{(i)} &= u^{(i)} + \xi e \quad \text{and} \\ \bar{u}^{(j)} &= u^{(j)} - \xi e \,, \end{aligned}$$

where ξ can be determined from the condition of the conservation of energy

$$(1.8.5) \qquad E(\bar{u}^{(i)}) + E(\bar{u}^{(j)}) = E(u^{(i)}) + E(u^{(j)}) \,,$$

when energy is presented here as a function of momentum. Calculating ξ now, we assume that $i \in K_1$ and $j \in K_2$.

From

$$\begin{aligned} E(\bar{v}^{(i)}) + E(\bar{u}^{(j)}) &= \frac{1}{2m_1} \left\langle u^{(i)} + \xi e, u^{(i)} + \xi e \right\rangle \\ &\quad + \frac{1}{2m_2} \left\langle u^{(j)} - \xi e, u^{(j)} - \xi e \right\rangle \end{aligned}$$

we get

$$(1.8.6) \qquad \xi = -2 \, \frac{\left\langle v^{(i)} - v^{(j)}, e \right\rangle}{\frac{1}{m_1} + \frac{1}{m_2}} \,.$$

Since the molecules i and j are approaching each other before the impact, we have

$$(1.8.7) \qquad \left\langle v^{(i)} - v^{(j)}, e \right\rangle < 0 \,.$$

From (1.8.6) and (1.8.7) we get

$$(1.8.8) \qquad \begin{aligned} \left\langle \bar{v}^{(i)} - \bar{v}^{(j)}, e \right\rangle &= \left\langle \frac{\bar{u}^{(i)}}{m_1} - \frac{\bar{u}^{(j)}}{m_2}, e \right\rangle \\ &= \left\langle \frac{1}{m_1}(u^{(i)} + \xi e) - \frac{1}{m_2}(u^{(j)} - \xi e), e \right\rangle \\ &= \left\langle v^{(i)} + \frac{\xi}{m_1} e - v^{(j)} + \frac{\xi}{m_2} e_1, e \right\rangle \\ &= \left\langle v^{(i)} - v^{(j)}, e \right\rangle + \xi \left(\frac{1}{m_1} + \frac{1}{m_2} \right) \\ &= - \left\langle v^{(i)} - v^{(j)}, e \right\rangle > 0 \,. \end{aligned}$$

Thus the molecules always move away from each other again. A dynamics with this property is called **separating**; see also Section 7.3 'Separating Dynamics'.

According to Theorem 4.2.1 of Cercignani et al. (1994) the iterative application of the described procedure yields a trajectory $(x(t); v(t))_{t \in \mathbb{R}_+}$ within $B^N \times \mathbb{R}^{2N}$ for Lebesgue – almost all starting points $(x(0), v(0)) \in B^N \times \mathbb{R}^{2N}$.

1.9 Conclusions

Experiment E 1.1 confirms part (a) of the Maxwell Hypothesis, cf. (1.1.1) and (1.5.7), for the velocity and also the momentum distribution. The running kernel density estimates of the projected velocity (momentum) distributions of a molecule onto L_φ, $0° \leq \varphi \leq 360°$, stabilize in the course of time, i. e. they are ergodic and they all coincide with the same centered normal distribution, its variance being estimated within the class of centered normal distributions based on data projected onto L_φ.

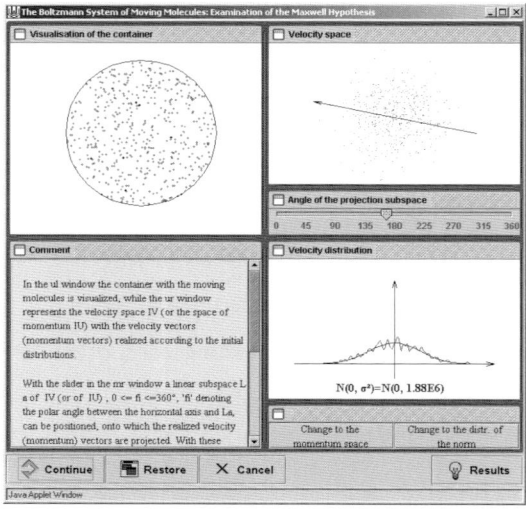

E 1.1. Examination of the Maxwell Hypothesis

At a level of significance of 0.05 we performed a χ^2-goodness-fit test, for several values of the polar angle φ, which did not reject the hypothesis that the kernel density estimates coincide statistically with the corresponding estimates within the class of centered normal distributions.

From this we conclude first that there exists an ergodic velocity (momentum) distribution on the velocity (momentum) space, and then that this ergodic velocity (momentum) distribution of a molecule is a centered, rotationally symmetric normal distribution, being the product measure of any pair of distributions projected onto L_φ, $0° \leq \varphi \leq 360°$.

2 Equilibrium Distribution and Entropy

2.1 Boltzmann–Gibbs Entropy

Let $(\Omega, \mathcal{A}, \mu)$ be a σ-finite measure space. According to the Radon–Nikodym theorem every μ-dominated probability measure P on $(\Omega, \mathcal{A})$ has a representation by a density function $f\colon \Omega \to \bar{\mathbb{R}}_+$ w.r.t μ, where $\bar{\mathbb{R}}_+ := \mathbb{R}_+ \cup \{+\infty\}$, cf. Bauer (2001), Theorem 17.10.

Lemma 2.1.1
For the μ-dominated probability measures P and Q on $(\Omega, \mathcal{A})$ with μ-densities f and g, respectively, we have

$$(2.1.1) \qquad - \int_\Omega f \log f \, \mathrm{d}\mu \ \leq \ - \int_\Omega f \log g \, \mathrm{d}\mu \,.$$

For a proof the reader is referred to Mackey (1992), p. 11.

Definition 2.1.2
Let P be a μ-dominated probability measure on $(\Omega, \mathcal{A})$ with a μ-density f. Then the value

$$(2.1.2) \qquad S(P) := \ - \int f \log f \, \mathrm{d}\mu$$

is called the **Boltzmann–Gibbs entropy** of P, whenever the integral involved is well–defined.

The name Boltzmann–Gibbs entropy honors the American Josiah Willard Gibbs (11 Feb 1839–28 April 1903) for his contributions to Statistical Mechanics.

Theorem 2.1.3 (cf. Mackey (1992), p. 11)
Let $H\colon \Omega \to \bar{\mathbb{R}}_+$ be a Borel-measurable function and let $\beta > 0$ such that

$$(2.1.3) \qquad Z := \int_\Omega \exp(-\beta \, H(\omega)) \, \mathrm{d}\mu(\omega) \ < \ \infty$$

and

$$(2.1.4) \qquad \varepsilon := \frac{1}{Z} \int_{\Omega} H(\omega) \exp(-\beta H(\omega)) \, d\mu(\omega) < \infty.$$

Then the probability measure P with the μ-density

$$(2.1.5) \qquad f := \frac{1}{Z} \exp(-\beta H)$$

is the unique optimizer of the Boltzmann–Gibbs entropy, i.e. $S(Q) \leq S(P)$ for all μ-dominated probability measures Q on $(\Omega, \mathcal{A})$ satisfying

$$(2.1.6) \qquad \int_{\Omega} H \, dQ = \varepsilon.$$

Remark 2.1.4
Instead of $(\Omega, \mathcal{A})$, we especially consider – with some modifications in Chapter 11 – the measurable space of momentum $\left(\underset{j=1}{\overset{N}{\times}} \mathbb{U}^{(j)}, \underset{j=1}{\overset{N}{\bigotimes}} U^{(j)} \right) = (\mathbb{R}^{dN}, \mathcal{B}^{dN})$ of N moving molecules in $\mathbb{R}^d$, $d = 2, 3$, while the function H, interpreted as the total kinetic energy, the so-called **Hamiltonian, is a sum of N identical terms**; we say that the Hamiltonian is **additively representable**. All the various Hamiltonians discussed in the sequel, especially in Chapters 7–12, are additively representable. The additive representability of the Hamiltonian together with the specification of the dominating measure μ in the definition of the Boltzmann–Gibbs entropy appear – from a mathematical point of view – as inherent principles of physics. Apart from Chapter 11, where μ is taken as the counting measure on a discrete space, we take μ to be a power of the Borel–Lebesgue measure λ on Borel σ-fields.

2.2 Maxwell Hypothesis and Entropy

We consider here the Boltzmann system of N moving molecules; in consequence we specify the measurable space $(\Omega, \mathcal{A})$ as introduced in Theorem 2.1.3 to be

$$(2.2.1) \qquad (\Omega, \mathcal{A}) := \left(\underset{j=1}{\overset{N}{\times}} \mathbb{U}^{(j)}, \underset{j=1}{\overset{N}{\bigotimes}} U^{(j)} \right) = (\mathbb{R}^{dN}, \mathcal{B}^{dN})$$

i.e., the measurable space of momentum of N molecules, cf. Section 1.5; while the measure μ introduced in Theorem 2.1.3 is now the Lebesgue measure λ^{dN} on $\mathcal{B}^{dN}$.

The Borel measurable function $H\colon \Omega \to \bar{\mathbb{R}}_+$ of Theorem 2.1.3 is determined as

(2.2.2)

$$H\colon \underset{j=1}{\overset{N}{\times}} \mathbb{U}^{(j)} \longrightarrow \mathbb{R}_+ \qquad \text{with}$$

$$H(u^{(1)}, \ldots, u^{(N)}) := \sum_{j=1}^{N} \frac{1}{2m} \langle u^{(j)}, u^{(j)} \rangle, \qquad u_j \in \mathbb{U}^{(j)}, j \in \mathbb{N}_N,$$

representing the Newtonian total kinetic energy of N molecules of mass $m > 0$ as function of their momenta.

For $\beta > 0$ the functions $Z(\beta)$ and $\epsilon(\beta)$ introduced in Theorem 2.1.3 become

(2.2.3)
$$Z(\beta) = \int_{\mathbb{R}^{dN}} \exp(-\beta H(u)) \, \mathrm{d}\lambda^{dN} = (2\pi m/\beta)^{dN/2}$$

and

(2.2.4)
$$\epsilon(\beta) = \frac{1}{Z(\beta)} \int_{\mathbb{R}^{dN}} H(u) \exp(-\beta H(u)) \, \mathrm{d}\lambda^{dN} = \frac{Nd}{2\beta} < \infty.$$

The density

$$f_\beta\colon \underset{j=1}{\overset{N}{\times}} \mathbb{U}^{(j)} = \mathbb{R}^{dN} \longrightarrow \mathbb{R}_+ \qquad \text{with}$$

(2.2.5)

$$f_\beta(u^{(1)}, \ldots, u^{(N)}) = \frac{1}{Z(\beta)} \exp(-\beta H(u^{(1)}, \ldots, u^{(N)}))$$

$$= \frac{1}{Z(\beta)} \exp\left(-\frac{\beta}{2m} \sum_{j=1}^{N} \langle u^{(j)}, u^{(j)} \rangle\right)$$

$$= \prod_{j=1}^{N} (2\pi m/\beta)^{-d/2} \exp\left(-\frac{\beta}{2m} \langle u^{(j)}, u^{(j)} \rangle\right)$$

determines a probability measure P_β on $\mathcal{B}^{dN}$:

(2.2.6)
$$P_\beta := f_\beta \cdot \lambda^{dN} = N(0, m/\beta \, I_d)^N = N(0, m/\beta)^{dN}.$$

According to Theorem 2.1.3, P_β is the optimizer of the Boltzmann–Gibbs entropy with respect to all λ^{dN} dominated probability measures Q on $\mathcal{B}^{dN}$ satisfying $\int H \, \mathrm{d}Q = \epsilon(\beta)$; the factor $N(0, m/\beta)$ of P_β being the distribution of a momentum component of a molecule.

Remark 2.2.1

For $\beta = 1/(k_\mathrm{B} \cdot T)$ the factor $N(0, m/\beta)$ of P_β becomes $N(0, k_\mathrm{B} \cdot m \cdot T)$, which is the distribution of a momentum component of a molecule according to the Maxwell Hypothesis for momenta, cf. (1.5.1).

2.3 Entropy Principle

Let a moving system of N molecules be given according to the laws of energy and momentum conservation, in which the forms of energy and momentum remain unspecified for the moment. Various forms will be investigated in Chapters 7–11 and 12, respectively, not necessarily having a real counterpart.

Let – apart from some straightforward modifications in Chapter 11 –

$$(2.3.1) \qquad (\mathbb{U}^{(j)}, \mathcal{U}^{(j)}) := (\mathbb{R}^d, \mathcal{B}^d) =: (\mathbb{U}, \mathcal{U}), \qquad j \in \mathbb{N}_N \,,$$

be the momentum space of the j-th molecule, while the Borel measurable mapping

$$(2.3.2) \qquad H_0 \colon \mathbb{R}^d \longrightarrow \mathbb{R}_+$$

represents the kinetic energy of a molecule as function of its momentum.

The total kinetic energy of the moving system of N molecules as function of their momentum, i.e. the Hamiltonian of this system, is given by the mapping

$$H \colon \mathbb{R}^{dN} \longrightarrow \mathbb{R}_+ \qquad \text{with}$$

$$(2.3.3) \qquad H(u) \;=\; \sum_{j=1}^{N} H_0(u^{(j)}) \,, \qquad u = (u^{(1)}, \dots, u^{(N)}) \in \mathbb{R}^{dN} \,,$$

for which we require

$$(2.3.4) \qquad Z_N(\beta) := \int_{\mathbb{R}^{dN}} \exp(-\beta H(u)) d\lambda^{dN}(u) \;<\; \infty$$

for $\beta > 0$: or equivalently

$$(2.3.5) \qquad Z_1(\beta) \;=\; \int_{\mathbb{R}^d} \exp(-\beta H_0(w)) d\lambda^d(w) \;<\; \infty \,.$$

Obviously we have by (2.3.3)

$$(2.3.6) \qquad Z_N(\beta) \;=\; (Z_1(\beta))^N \,.$$

The density

$$(2.3.7) \qquad f_\beta \colon \mathop{\vcenter{\hbox{$\times$}}}_{j=1}^{N} \mathbb{U}^{(j)} \;=\; \mathbb{R}^{dN} \longrightarrow \mathbb{R}_+$$

with

$$f_\beta(u^{(1)}, \ldots, u^{(N)}) := \frac{1}{Z_n(\beta)} \exp\Big(-\beta H(u^{(1)}, \ldots, u^{(N)}) \Big)$$

(2.3.8)

$$= \prod_{j=1}^{N} \frac{1}{Z_1(\beta)} \exp(-\beta H_0(u^{(j)})) =: \prod_{j=1}^{n} (f_\beta^0(u^{(j)}))$$

determines a probability measure P_β on $\mathcal{B}^{dN}$, being the N-th power of a probability measure P_β^0 on $\mathcal{B}^d$:

(2.3.9) $$P_\beta = f_\beta \lambda^{dN} = (f_\beta^0 \lambda^d)^N = (P_\beta^0)^N.$$

2.3.1 Definition

Let $\beta > 0$ and let H_0 and H be as in (2.3.2) and (2.3.3). Then the family of **entropic momentum distributions of a molecule** and **of the system of N moving molecules** is defined by

(2.3.10) $$\{P_\beta^0 | \beta > 0\}$$

and

(2.3.11) $$\{P_\beta | \beta > 0\},$$

respectively. Instead of P_β^0 and P_β we also write $\boldsymbol{P_{\mathrm{ent}}^0(\beta)}$ and $\boldsymbol{P_{\mathrm{ent}}(\beta)}$.

Note that $P_{\mathrm{ent}}(\beta)$ is the entropy optimizer according to Theorem 2.1.3 for H as specified by (2.3.3).

On the other hand we have the **empiric momentum distribution** P_{erg}^0 and P_{erg} as a physically or computer experimentally realized momentum distribution of a molecule and of the whole system of moving molecules, respectively. We will refer to the empiric momentum distribution **in an equilibrium state** as the **ergodic distribution** for short, which of course requires first proving experimentally that the **empiric distribution stabilizes**.

The computer experimentally realized ergodic distributions are approximated by their density estimates based on the experiments with moving molecules.

With these definitions the **Entropy Principle** is formulated, which has to be seen as a **generalized Maxwell Hypothesis**.

2.3.2 Entropy Principle

Let a moving system of N molecules be given according to the laws of momentum and energy conservation; its Hamiltonian, being additively representable, fulfills the conditions (2.3.4) or equivalently (2.3.5) for $\beta > 0$.

(a) There exists a $\beta^* > 0$, such that

(2.3.12) $$P_{\mathrm{erg}}^0 = P_{\mathrm{ent}}^0(\beta^*).$$

(b) The temperature T^* of the moving system of molecules is given (defined) by

(2.3.13)
$$T^* = 1\big/(k_\mathrm{B} \cdot \beta^*).$$

We are speaking of β^* and T^* as the *accurate* parameter β^* and the *accurate* temperature T^*, respectively.

One notes that the entropic distributions are based only on the Hamiltonian H, i. e. on the total kinetic energy, while the empirically realized (ergodic) distribution requires the validity of the laws of energy and momentum conservation and the specification of the mathematical forms of energy and momentum.

For the case of Newtonian dynamics with the kinetic energy being given by (2.2.2) and the momentum of a molecule by $u^{(j)} = mv^{(j)}$, $j \in \mathbb{N}_N$, we know from Remark 2.2.1 and experiment E 1.1 that the Entropy Principle is valid, i. e. it is equivalent to the Maxwell Hypothesis for momenta, cf. Principle 1.5.1.

In Chapters 7–12 we investigate the Entropy Principle for various forms of energy and momentum, which do not necessarily have a counterpart in real physics.

Thereby of course the distributional aspect of the Entropy Principle, i. e. its statement (a), may be validated, while the more intrinsic physical aspect, i. e. statement (b), can not be tested on the computer not even for a Boltzmann system of moving molecules subject to Newtonian dynamics. Statement (b) yields a definition of temperature, based on statement (a), ensuring that β^* and therefore also T^* are scalar quantities, which indeed can be tested in the computer experiments.

Note that Max Planck had generalized the Maxwell Hypothesis substituting the special term of the kinetic energy in the exponent of the density of the Maxwell–Boltzmann distribution by the general energy function.

3 Measurements in the Boltzmann System

Various phenomena of the Boltzmann system of moving molecules are investigated, in particular the collision (reflection) process, those events are defined by the collision of arbitrary pairs of a molecules (or of an arbitrary molecule with a segment of the boundary) as well as – in contrast to this – the collision of certain molecules along its trajectory with other molecules (or with a segment of the boundary).

3.1 Global Collision Process

Problem

Let us again consider the Boltzmann system of moving (virtual) molecules confined to the 2-dimensional container as mentioned in previous sections.

The intention now is to examine the global collision process, i.e. the objects of our interest are the collisions between **arbitrary pairs** of molecules, cf. experiment E 3.1, as well as the reflections of arbitrary molecules with a specified segment of the boundary of the container, cf. experiment E 3.2. The aim is to specify the counting process defined by the collisions and the hits, respectively.

The Poisson Process and the Distribution of its Inter-Event Times

Let $(\Omega, \mathcal{A}, P)$ be a probability space and (T_i) an independent sequence of identically distributed real random variables. Let (S_m) denote the sequence of the partial sums of the T_i:

(3.1.1)
$$S_m := \sum_{i=1}^{m} T_i, \qquad m \in \mathbb{N}.$$

Then the sequence (S_m) induces a counting process $(N_t)_{t \geq 0}$ with

(3.1.2)
$$N_t := \#\{i \in \mathbb{N} \mid S_i \leq t\},$$

where t is an arbitrary point of the time axis and $\#S$ stands for the cardinality of the set S.

This process is called the Poisson process Π_γ with intensity γ, iff the increments

$$N_{t_2} - N_{t_1}, \qquad 0 \le t_1 < t_2 \le \infty,$$

are distributed according to the Poisson distribution π_β with parameter $\beta := (t_2 - t_1)\gamma$, and probability function w_β given by

$$w_\beta(k) := e^{-\beta} \frac{\beta^k}{k!}, \qquad k \in \mathbb{N}.$$

From this it follows that

$$(3.1.3) \qquad \mathbb{E}(N_{t_2} - N_{t_1}) = (t_2 - t_1) \cdot \gamma, \qquad 0 \le t_1 < t_2 < \infty,$$

which means that the expectation of the increment $N_{t_2} - N_{t_1}$ depends only on the length of the time interval $[t_1, t_2)$ and not on its location on the time axis.

A Poisson process can be characterized as a counting process with stochastically independent and stationary increments, for which the events do not accumulate at any time point.

Calling the quantities $T_i := S_i - S_{i-1}$ the inter-event times, $i \in \mathbb{N}$, the Poisson process Π_γ is determined by an independent sequence of random variables inter-event times T_i being distributed according to the exponential distribution $\mathrm{Exp}(\gamma)$ with parameter γ.

A Lebesgue density f_γ of $\mathrm{Exp}(\gamma)$ is given by

$$(3.1.4) \qquad f_\gamma(x) = \begin{cases} \gamma \exp(-\gamma x) & \text{for } x \ge 0 \\ 0 & \text{elsewhere}. \end{cases}$$

Examining the Poissonity of the Collision/Reflection Processes

The collision/reflection processes of both experiments E 3.1 and E 3.2 enjoy – at least from an intuitive viewpoint – the properties characterizing a Poisson process mentioned above.

In order to check the Poissonity of the collision/reflection process in experiment E 3.1 as also in experiment E 3.2, we examine whether the inter-event times follow an exponential distribution. Classical goodness-of-fit tests are not applicable in this context, as neither the type of process nor the actual parameters are known.

To overcome this difficulty, the density of the distribution of the inter-event times is estimated with the kernel density estimator, which is a non-parametric procedure, delivering only numerically the graph of the required density. In addition, the kernel density estimate will be compared graphically with the graph of the density of $\mathrm{Exp}(\gamma)$, γ being estimated parametrically within the class of the exponential distributions.

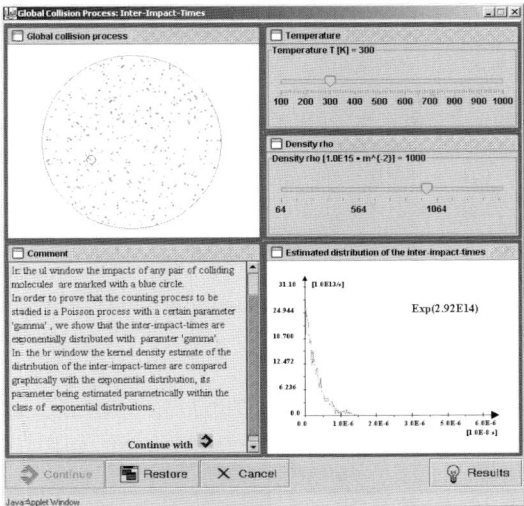

E 3.1. Global inter-impact times

For $x_i \in \mathbb{R}$, $i \in \mathbb{N}$, being realizations of $\mathrm{Exp}(\gamma)$ a consistent estimator $\hat{\gamma}$ of parameter γ is given by

$$(3.1.5) \qquad \hat{\gamma} = \left(\frac{1}{n} \sum_{i=1}^{n} x_i \right)^{-1} .$$

Estimating the parameter γ of the exponential distribution $\mathrm{Exp}(\gamma)$ within experiment E 3.2, we count the hits of arbitrary molecules with the boundary within a certain segment ('detector'); thereby we learn that the estimate of γ does not depend on the position of the 'detector' within the container.

To show, that the inter-event times can be seen as realizations of an independent sequence of random variables, the so-called permutation test may be applied for instance, cf. Moeschlin et al. (2003), p. 30.

Comments
By the experiments E 3.1 and E 3.2 the exponentiality of the inter-event times is confirmed (which is a necessary condition for the Poissonity of the respective collision/reflection processes). Note that the actual parameters of the two processes considered in E 3.1 and E 3.2 do not coincide.

No influence of the positioning of the specified segment on the distribution parameter γ of the inter-hit times in experiment E 3.2 could be ascertained.

This result is remarkable insofar as the parameter γ enters into the Formula (5.1.12) of the asymptotic value of the estimator for the pressure. Indeed, the pressure of a fluid within a container also does not depend on the location of the manometer (detector) within the container.

3.2 Collision Process along a Trajectory

Problem

Again, the Boltzmann system of moving (virtual) molecules confined to a 2-dimensional container is considered.

The intention now is to investigate the inter-impact times, cf. E 3.3, and the inter-impact lengths (free path lengths), cf. 3.4, **along a molecular trajectory**. These inter-impact times and path lengths, can respectively be interpreted as survival data, which in the 'non-aging case' usually are modeled by exponential distributions. We follow this line here, checking the exponentiality of these inter-impact data.

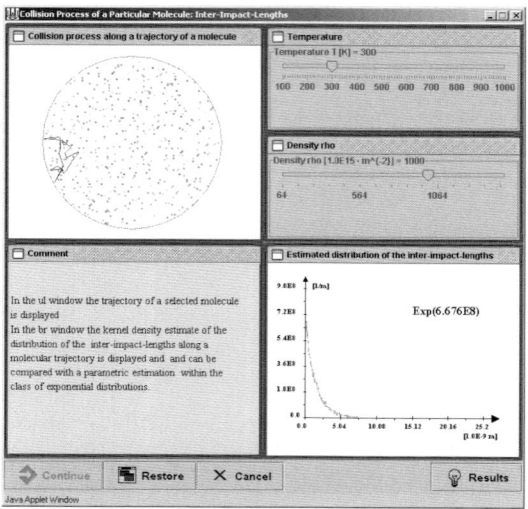

E 3.4. Inter-impact lengths along a trajectory

Examining the Exponentiality of the Inter-impact Data

The technique of estimation is the same as described in 3.1.

Comments

By means of experiments E 3.3 and E 3.4, the exponentiality of the inter-impact times and inter-impact path lengths, respectively, is confirmed. The actual parameters of the respective exponential distribution carry different physical units; they do not coincide.

3.3 Reflection Angles

Problem

For the Boltzmann system of moving (virtual) molecules confined to a 2-dimensional container we explore now the distribution of the reflection angles of those molecules whose the next collision is a hit with a specified 'detection' segment of the boundary.

Realizations of small reflection angles are privileged since, for a molecule with a fixed position, a small angle means a shorter path to the boundary, thus increasing the probability for the next collision to be a hit with the boundary.

Within experiment E 3.5 various shapes of a triangular container are studied, where the localization of the detection segment can be specified by the user.

Exploring the Distribution of the Reflection Angles

The distribution of the reflection angles is estimated by the kernel density estimator.

Comments

Our prognosis that small reflection angles are privileged proves to be true.

Astonishingly enough, the distribution of the reflection angles is described by the Lebesgue density

$$f(\alpha) \;=\; \cos(\alpha)\,, \qquad 0 \le \alpha \le \pi/2\,,$$

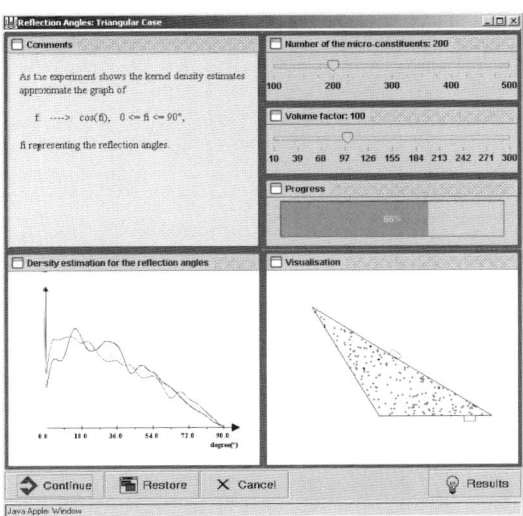

E 3.5. Reflection angles

for any shape of the container and for any position of the 'detection' segment; a fact which can be proved only for some special case regarding the shape of the container.

One may ask whether there is an analogous result for 3-dimensional systems. The statistical evaluation of corresponding computer experiments suggests that in the 3-dimensional case there is also a universal distribution of the reflection angles whose Lebesgue density is given by

$$f(\alpha) \;=\; 2\sin(2\alpha)\,, \qquad 0 \le \alpha \le \pi/2\,.$$

3.4 Collision Probabilities of Pairs of Molecules

Problem
For the Boltzmann system of moving (virtual) molecules confined to a 2-dimensional container we explore the collision probabilities of pairs of molecules.

Exploring the Distribution of Pairs of Colliding Molecules
By an averaging process with respect to the relative frequencies of each pair of molecules, the respective probabilities are estimated based on the validity of the strong law of large numbers.

Comments
The distribution of pairs of colliding molecules is the uniform distribution on the Cartesian product of the set of molecules with itself. In other words, the

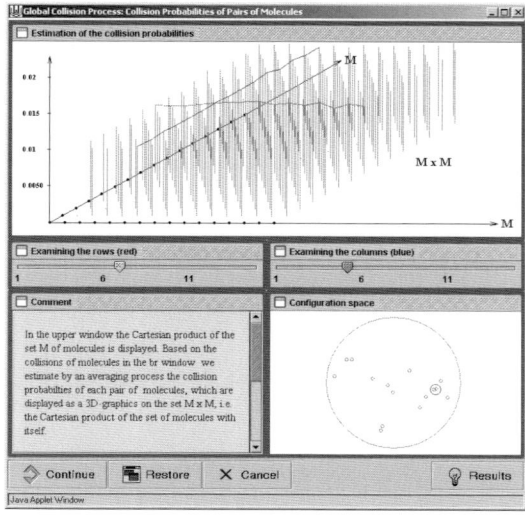

E 3.6. Collision probabilities of pairs of molecules

Boltzmann system of moving molecules ensures that each pair of molecules enjoys the same probability that they collide.

The consequence of the results of the present experiment E 3.6 as well as of experiment E 3.1, stating that the inter-collision times are distributed according to $Exp(\gamma)$, allows us to substitute the collision model of Ludwig Boltzmann (Boltzmann system) by a stochastical model of energy splitting which is no longer based on collisions, cf. Chapters 9 and 11.

4 Average Free Path Length

4.1 Examining a Product Formula

Average free path length is a well known notion from Statistical Mechanics, which prompts us to investigate its relations to the average inter-impact time.

Problem
In the sequel we examine the formula:

(4.1.1)   average free path length
$$= \text{average inter-impact time} \times \text{average velocity},$$

which will be examined in experiment E 4.1.

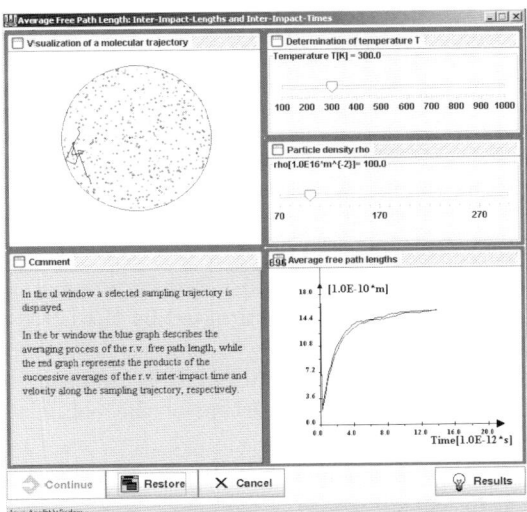

E 4.1. Average free path length

Method of Examination
We compare graphically the successive averages of a realized sequence of the free path lengths in its temporal evolution, with the product of the respective averages of the inter-collision times and of the velocities, respectively.

Comments
The trajectories of the estimation processes become stabilized for a great number of sample values. There is a near coincidence of the right- and left side term of Equation (4.1.1), but not a precise one.

4.2 Measuring the Correlations

Problem
The result of experiment E 4.1 gives rise to study the interrelationship between the sequencies of the random variables free path lengths, inter-impact times and velocities. In the sequel we examine their correlatedness.

Experiment
In experiment E 4.2 we check the lack of correlation of any pair of the involved variables, estimating the respective correlation coefficients.

Comments
The sequencies of the variables free path lengths (inter-impact lengths), inter-impact times and velocities are not uncorrelated and therefore not stochastically independent.

4.3 The Effect of the Particle Density

Problem
In experiment E 4.3 the question of the dependence of the free path length on the particle density of the fluid is addressed.

Examination Method
The method of analysis is straightforward. For various particle densities the expectation of the free path length is estimated by the corresponding successive averages.

The stabilized estimates are transferred to a diagram representing the expectation of free path lengths as function of the particle densitiy, cf. experiment E 4.3.

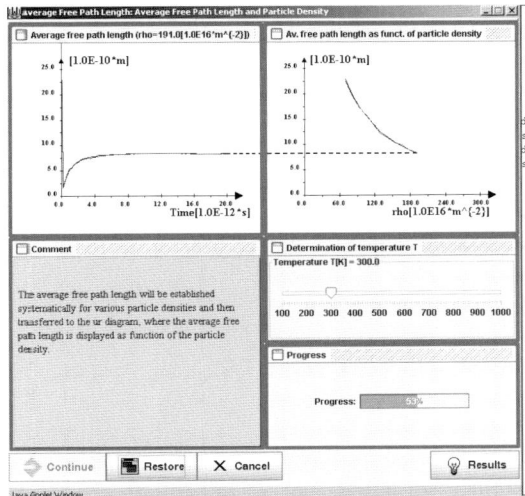

E 4.3. Free path length versus particle density

Comments

As supposed, a higher particle density implies shorter path lengths. For at least a certain part of the particle density axis, the free path lengths are proportional to the inverse of the particle density; cf. Jäckle (1978).

5 Estimation of Pressure

Based on the physical definition of pressure we define an estimator for the average pressure during a time period τ on a segment A of the boundary.

From a formula for the value of the asymptotic pressure, we get a relation to the parameter γ of the collision (hit) process of arbitrary molecules with a segment of the boundary of the container; E 3.2.

Important for physical reality is the fact that the pressure within a container of moving molecules according to the Boltzmann system does not depend on the position of the segment on which the pressure is measured.

An application for the measuring of pressure of virtual molecules delivers the theory of osmotic pressure with the well-known law of van't Hoff, which is examined on the computer.

5.1 Pressure Estimation

In the kinetic gas theory the notion of pressure plays an important role. To develop a tool for measuring the pressure of a virtual fluid in a container, i. e. to develop an estimator for the pressure of a Boltzmann system of moving molecules within a container, we have to give an account how pressure is defined.

Let us again consider a 2-dimensional container with molecules on which Newtonian dynamics is imposed. When molecules strike a segment A of the boundary having length $\lambda^1(A)$ under the Lebesgue measure λ^1, the components U_j of the momentum orthogonal to the boundary change from U_j to $-U_j$. Its difference $2U_j$ has to be summed over all molecules striking the boundary within the segment A during a certain time period. Dividing the value of this sum by the length of the time period and by $\lambda^1(A)$, yields the average pressure on A during the considered time period.

Let $(\Omega, \mathcal{A}, P)$ be a probability space and (U_j) an independent sequence of quadratically integrable random variables U_j modeling the momenta orthogonal to the boundary of the j-th molecule striking the boundary within A.

The expectation $\mathbb{E}(U_j)$ of the identically distributed random variables U_j will be denoted by U, $j \in \mathbb{N}$.

The sequence (S_m) of random variables S_m, modeling the time points at which a molecule is striking the boundary within A, is given by

$$(5.1.1) \qquad\qquad S_m := \sum_{i=1}^{m} T_i \,,$$

where the independent sequence (T_i) of quadratically integrable random variables T_i are modeling the time periods between two successive molecules striking the boundary within the fixed segment A.

From experiment E 3.2 we know that the random variables T_i, $i \in \mathbb{N}$, are identically distributed according to the exponential distribution $\mathrm{Exp}(\gamma)$ with parameter $\gamma > 0$. Thus we have by a well-known fact about the expectation of exponential distributed random variables

$$(5.1.2) \qquad\qquad \gamma = 1/\mathbb{E}(T_i) \,, \qquad i \in \mathbb{N} \,.$$

With these preparations we define an estimator $\hat{P}_{S_m}$ for the average pressure on the segment A during the time period defined by S_m:

$$(5.1.3) \qquad\qquad \hat{P}_{S_m} := \frac{2 \sum\limits_{j=1}^{m} U_j}{S_m \cdot \lambda^1(A)} \,.$$

Theorem 5.1.1.
The sequence $(\hat{P}_{S_m})$ of estimators $\hat{P}_{S_m}$ for the average pressure on A during the time interval determined by $S_m = \sum\limits_{i=1}^{m} T_i$ is strongly consistent:

$$(5.1.4) \qquad\qquad \lim_{m\to\infty} \hat{P}_{S_m} = \frac{2\gamma U}{\lambda^1(A)} \qquad P\text{-a.e.}$$

Proof:
By the assumptions on the sequences (U_j) and (T_i), the strong law of large numbers is valid for both sequences, and so we get from (5.1.2).

$$\lim_{m\to\infty} \hat{P}_{S_m} = \lim_{m\to\infty} \frac{2 \sum\limits_{j=1}^{m} U_j}{\sum\limits_{i=1}^{m} T_i \cdot \lambda^1(A)} = \lim_{m\to\infty} \frac{2 \cdot \frac{1}{m} \sum\limits_{j=1}^{m} U_j}{\frac{1}{m} \sum\limits_{i=1}^{m} T_i \cdot \lambda^1(A)}$$

$$= \frac{2U}{\gamma^{-1} \cdot \lambda^1(A)} = \frac{2\gamma U}{\lambda^1(A)} \qquad P\text{-a.e.} \qquad \square$$

Now the time interval during which the average pressure is measured will no longer be determined by S_m but by a fixed time length τ, with the consequence that the number of hits on the segment A during the time length τ is also a random variable.

Let (τ_n), $0 < \tau_n \in \mathbb{R}$ with $\tau_n \to \infty$ for $n \to \infty$, be a sequence of times during which the pressure on the segment A is measured, i. e. is estimated.

For any $\tau_n > 0$ the random variable

$$(5.1.5) \qquad\qquad M_{\tau_n} : (\Omega, \mathcal{A}, P) \;\to\; (\mathbb{N}, \mathcal{P}(\mathbb{N}))$$

defined by

$$(5.1.6) \qquad M_{\tau_n}(\omega) := \max\{j \in \mathbb{N}\,|\, S_j(\omega) \le \tau_n\}, \quad \omega \in \Omega,$$

counts the number of hits on A during τ_n.

This gives rise to defining the estimator $\hat{P}_{\tau_n}$ of the average pressure on A during the time period τ_n by

$$(5.1.7) \qquad\qquad \hat{P}_{\tau_n} := \frac{2 \displaystyle\sum_{j=1}^{M_{\tau_n}} U_j}{\tau_n \cdot \lambda^1(A)},$$

its number of hits M_{τ_n} being randomly defined.

From the definition of S_m, cf. (5.1.1), it follows that

$$(5.1.8) \qquad\qquad M_{\tau_n} \to \infty \quad P - \text{a.e. for } n \to \infty.$$

The sequence $\left(S_{M_{\tau_n}(\omega)}(\omega)\right)$ with elements

$$(5.1.9) \qquad\qquad S_{M_{\tau_n}(\omega)}(\omega) \;=\; \sum_{j=1}^{M_{\tau_n}(\omega)} T_j(\omega)$$

is therefore an infinite subsequence of $(S_m(\omega))$ with

$$(5.1.10) \qquad S_{M_{\tau_n}(\omega)}(\omega) \;\le\; \tau_n \;<\; S_{M_{\tau_n}(\omega)+1} \quad \omega \in \Omega, \;\; n \in \mathbb{N},$$

such that

$$(5.1.11) \qquad\qquad \lim_{n\to\infty} \hat{P}_{S_m} \;=\; \lim_{n\to\infty} \hat{P}_{S_{M_{\tau_n}}} \;=\; \lim_{n\to\infty} \hat{P}_{\tau_n}.$$

Theorem 5.1.2.
The sequence $\left(\hat{P}_{\tau_n}\right)$ of estimators $\hat{P}_{\tau_n}$ for the average pressure on A during the time period τ_n is strongly consistent:

$$(5.1.12) \qquad\qquad \lim_{n\to\infty} \hat{P}_{\tau_n} \;=\; \frac{2\gamma u}{\lambda^1(A)}.$$

Proof:
(5.1.12) follows by Theorem 5.1.1 and (5.1.11).

Theorem 5.1.2 justifies the estimator $\hat{P}_{\tau_n}$ for large τ_n as a reliable estimator for the average pressure on A during the time period τ_n.

Problem
The important question about the pressure is: "Does the average pressure on A during a certain time period depend on the positioning of A?" An answer is given by experiment E 5.1.1.

Experiment
In experiment 5.1.1 it is experimentally confirmed that the pressure on segment A does not depend on the position of A. The comparison of pressure estimates shows no significant differences.

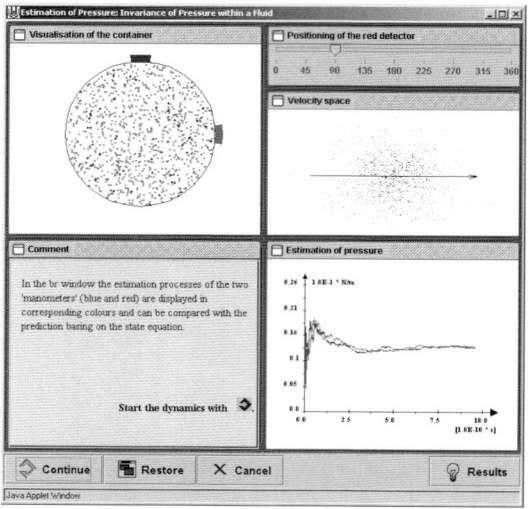

E 5.1. Invariance of pressure

The invariance of pressure within a fluid being postulated in Thermodynamics and having numerous applications in hydraulics and pneumatics turns out to be a consequence of the Newtonian dynamics imposed on a Boltzmann system of moving molecules.

5.2 Osmotic Pressure

The Boltzmann system of moving molecules can be applied to the investigation of phenomena from physical chemistry, as e.g. the study of the diffusion process. The subject of our investigations is van't Hoff's formula, i.e. van't Hoff's law.

Soluble and Solvent

Let a rectangular container be divided by a membrane into two congruent, rectangular cells.

Cell 1 is initially filled with the soluble; its molecules having mass $m_1 > 0$; cell 2 is filled with the molecules of the solvent having mass $m_2 > 0$. Since Newtonian dynamics is imposed on both types of molecules the velocity distributions in both cells are given initially by $N\left(0_2, \frac{k_\mathrm{B} \cdot T_i}{m_i} I_2\right)$, $i = 1, 2$.

| Cell 1 | mass m_1 | Cell 2 | mass m_2 |

soluble: N_1 molecules   Membrane   solvent: N_2 molecules

A molecule of the soluble cannot pass the membrane, and its movements are restricted to cell 1. The molecules of the solvent, initially located in cell 2, may pass through the membrane, i.e. they can move within both cells, 1 and 2, with the consequence that some of the molecules of the solvent enter into cell 1, such that the pressure in the cells will differ. This difference of pressure is called the 'osmotic pressure'.

Van't Hoff's Law

Jacobus Hendricus van't Hoff (30 August 1852–01 March 1911) was awarded the first Nobel Prize for chemistry in 1901 for his discovery of osmotic pressure for dilute solutions and the so-called van't Hoff law, formulated for incompressible solvents.

Let V be the volume of the cell with the soluble (and the solvent); i.e. cell 1.

If the number N_2 of molecules of the solvent is large in comparison with the number N_1 of molecules of the soluble, i.e. if the solution is diluted, then the osmotic pressure is given – according to van't Hoff's formula – by

$$\mathrm{P_{osm}} \approx \frac{N_1 \cdot k_\mathrm{B} \cdot T}{V}.$$

Problem

Within the framework of the arrangement described in 'soluble and solvent' the osmotic pressure is to be measured, i. e. estimated for $N_1 \ll N_2$.

Thereby one is interested in the number of molecules of the solvent present within cell 2 of the soluble at regular time points.

Moreover van't Hoff's law, which implies – within a certain range – the linearity of the osmotic pressure as function of the number of molecules of the soluble, will be established by a computer experiment.

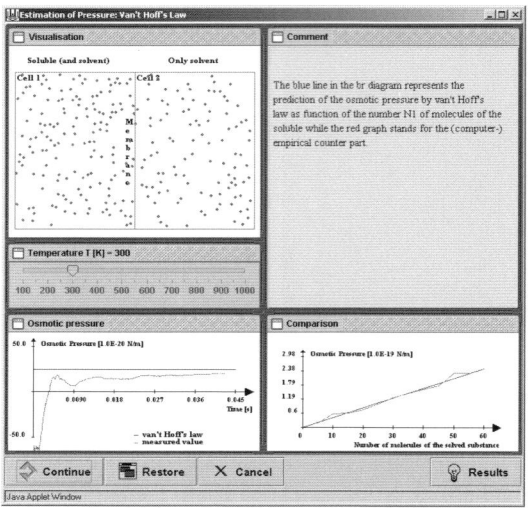

E 5.3. Examining van't Hoff's law

Experiment

For both experiments E 5.2 and E 5.3 we take as a basis the arrangement described in 'soluble and solvent' where $T_1 = T_2$ is prescribed. This can be reached for example by the Box–Müller procedure which yields realizations of the 2-dimensional normal distribution $N(0, \sigma^2) \otimes N(0, \sigma^2)$ with a prescribed variance σ^2, cf. Remark 1.4.1.

The dynamics of the molecules is the one described in 1.8, where two types of molecules are considered, having different masses m_i, $i = 1, 2$.

We base our estimation of osmotic pressure on the estimator (5.1.7).

In both experiments E 5.2 and E 5.3 the number N_2 of molecules of the solvent is prescribed. While the number N_1 of molecules of the soluble can be fixed by the experimenter in experiment E 5.2, this number is variable in experiment E 5.3; it changes automatically.

Experiment E 5.2 demonstrates the temporal evolution of the estimation process for the osmotic pressure.

Experiment E 5.3 displays the osmotic pressure for various numbers of molecules of the soluble in order to verify, i. e. to confirm, van't Hoff's law. E 5.3 is based on E 5.2 for alternative numbers N_1 of the molecules of the soluble.

6 An Example of Non-Real Physics

6.1 A Plea for Non-Real Physics

The Boltzmann system of moving molecules based on Newtonian dynamics shows astonishing phenomena, as for instance the ergodic distribution of momentum being the optimizer of the Boltzmann–Gibbs entropy, cf. Chapter 2, and yields together with the Maxwell Hypothesis deep insights relating temperature and energy.

How far do these experiences imply each other, respectively which of these fundamental concepts perceived in physical reality are independent of each other, and cannot be fathomed when moving only within physical reality?

What is special for our physical reality? Which phenomena remain valid even beyond this physical reality?

To approach such questions our analysis in this and also further chapters is based on models, in which obvious properties of real physics are no longer valid. Are the consequences of such alterations predictable? Which properties remain valid and which change?

A possible way to effect such explorations is offered by the virtual laboratory, i.e. the computer.

6.2 One-Way Permeable Membranes

Let us once more consider the Boltzmann system of N moving molecules, all with the same positive radius and the same positive mass m, confined to a ring-shaped container B, which is divided into two congruent parts by two one-way-permeable membranes, which can be passed by the molecules in only the mathematically positive sense. This, of course, contradicts nature.

This suggests the question, what are the consequences of these one-way-permeable membranes for the movement of the molecules. The intuitive answer that the molecules will enter into a rotational movement in the mathematical positive sense proves correct, cf. experiment E 6.1, in which the rotational velocity of the cloud of particles is estimated. Moreover we are able to confirm by experiment E 6.2 that the Maxwell Hypothesis still holds true, so that the notion of temperature can be introduced, cf. E 6.3.

6.3 Dynamics of the Experiment

The dynamics to be implemented on the computer is that as described in Section 1.8, of course with only one kind of molecules. The impacts of the molecules with the membranes from one side are treated in the same way as impacts of the molecules with the boundary of the container; while from the other side there are no impacts between molecules and membranes.

6.4 Estimating the Rotational Velocity

In experiment E 6.1 the rotational velocity of the molecules taken as a cloud is estimated. To this end we describe the momentary micro-state of the system by the $4N$-tuple

$$\left(x^{(1)}(t),\ldots,x^{(N)}(t);\ v^{(1)}(t),\ldots,v^{(N)}(t)\right),$$

$x^{(j)}(t)$ describing the positions and $v^{(j)}(t)$ the velocities of the molecules at time t.

The initial positions $x^{(j)}(0)$ and the initial velocities $v^{(j)}(0)$ are generated according to the uniform distribution on B and the normal distribution

$$N(0,\sigma^2(0)) \otimes N(0,\sigma^2(0))$$

on $\mathcal{B}^2$ with $\sigma^2(0) > 0$, respectively.

According to the generation of initial velocities, the mechanical energy of the system is 0 at time 0, while the thermal energy $E_{textth}(0)$ at time 0 is given by the total kinetic energy of the micro-constituents:

(6.4.1) $$E_{\text{th}}(0) \;=\; Nm\sigma^2(0)\,.$$

We introduce the tangential and the radial component of $v^{(j)}(t)$:

(6.4.2) $$v_{\text{T}}^{(j)}(t) \;:=\; \langle v^{(j)}(t), e_{\text{T}}(x^{(j)}(t))\rangle$$

and

(6.4.3) $$v_{\text{R}}^{(j)}(t) \;:=\; \langle v^{(j)}(t), e_{\text{R}}(x^{(j)}(t))\rangle\,,$$

$e_{\text{T}}(x^{(j)}(t))$ and $e_{\text{R}}(x^{(j)}(t))$ denoting the unit vector in the tangential and the radial direction at $x^{(j)}(t)$, respectively, $j = 1,\ldots,N$.

Experiment E 6.1 demonstrates that the system of molecules enters into a rotational movement in the mathematically positive sense with at first a strongly increasing angular velocity. Thereby the tangential velocity of a particle can be understood as the sum of the rotational velocity of the cloud (total system) cf molecules – i. e. the angular velocity of the cloud multiplied by the Euclidean norm $|x^{(j)}(t)|$ of the positions $x^{(j)}(t)$ of the molecules – and a remaining individual thermal, tangential velocity of the molecule, so that the angular velocity of the cloud of particles can be quantified solving a least-square ansatz:

$$(6.4.4) \qquad \sum_{j=1}^{N} \left(v_{\mathrm{T}}^{(j)}(t) \ - \ \omega(t) \cdot |x^{(j)}(t)| \right)^2 \ \longrightarrow \ \min$$

with respect to $\omega(t)$ at regular time intervals.

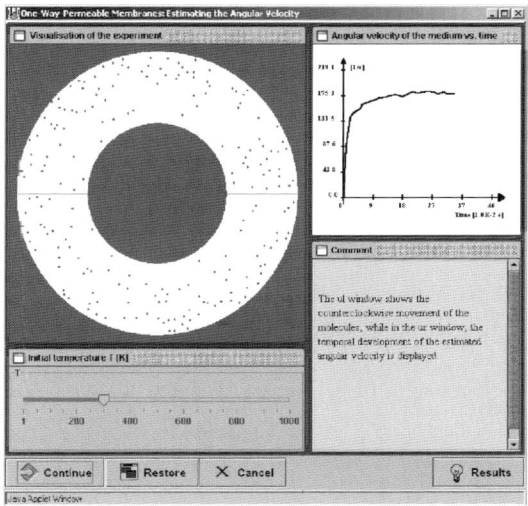

E 6.1. Angular velocity versus time

6.5 Examining the Maxwell Hypothesis

As experiment E 6.1 shows, our ring-shaped container with the two only one-way-permeable membranes is a thermodynamic machine which converts heat, i. e. disordered energy, into rotational kinetic energy of the fluid without having two pools cf heat.

The estimation of the angular velocity of the fluid, cf. E 6.1, allows the purely thermal velocity vectors of the molecules at a certain time point to be determined. The question is, whether the Maxwell Hypothesis, statement (a), cf. (1.1.1), still holds true for the remaining thermal energy.

Provided that the Maxwell Hypothesis remains valid even in this context, the thermal velocity vectors at time t

$$(6.5.1) \qquad \begin{pmatrix} v_{th_1}^{(j)}(t) \\ v_{th_2}^{(j)}(t) \end{pmatrix} = \begin{pmatrix} v_{\mathrm{T}}^{(j)}(t) - \omega(t) \cdot |x^{(t)}(t)| \\ v_{\mathrm{R}}^{(j)}(t) \end{pmatrix} , \quad j = 1, \dots, N,$$

are distributed according to

$$(6.5.2) \qquad \begin{cases} N(0, \sigma^2(t)) \otimes N(0, \sigma^2(t)) & \text{with} \\ \sigma^2(t) := \sigma^2(0) - \frac{1}{mN} E_{\mathrm{rot}}(t) & \text{and} \\ E_{\mathrm{rot}}(t) := \frac{1}{2} m \sum\limits_{j=1}^{N} \omega(t)^2 \cdot |x^{(j)}|^2 . \end{cases}$$

This distribution of the thermal velocity vectors will be examined and proved experimentally in E 6.2:

Let $\hat{\omega}(t)$ be the estimate of the angular velocity of the fluid at a time instant t according to (6.4.4). Then the (estimated) vectors $\hat{v}_{\mathrm{th}}^{(j)}$ of the thermal velocities are given by

$$\hat{v}_{\mathrm{th}}^{(j)} = \begin{pmatrix} v_{\mathrm{T}}^{(j)}(t) - \hat{\omega}(t) \cdot |x^{(j)}(t)| \\ v_{\mathrm{R}}^{(j)}(t) \end{pmatrix} , \quad j = 1, \dots, N,$$

and can be viewed as elements of an abstract 2-dimensional linear space in which the first component $\hat{v}_{\mathrm{th}1}^{(j)}$ of $\hat{v}_{\mathrm{th}}^{(j)}$ corresponds to the tangential direction and the second to the radial one.

Let φ be the polar angle of a linear subspace L_2 with unit vector

$$e_\varphi := \begin{pmatrix} \cos \varphi \\ \sin \varphi \end{pmatrix} .$$

Then the distribution of the projections $\langle \hat{v}_{\mathrm{th}}^{(j)}, e_\varphi \rangle$, $j = 1, \dots, N$, of the thermal velocity vectors onto L_2 can be estimated by a kernel density estimator.

The resulting estimates are then compared graphically with the corresponding entropic (normal) distribution

$$N(0, \sigma^2(t)) \otimes N(0, \sigma^2(t))$$

in analogy to the examination in E 1.1.

6.6 Temporal Evolution of Temperature

Having established with experiment E 6.2 the validity of the Maxwell Hypothesis even in the context of our ring-shaped non-real thermodynamic machine, we are in a position based on (1.4.3) to estimate the temperature at regular time intervals, which is subject of experiment E 6.3. Indeed, this temperature decreases in the course of time according to the fact that the kinetic energy of the molecular system (fluid) increases, cf. experiment E 6.1.

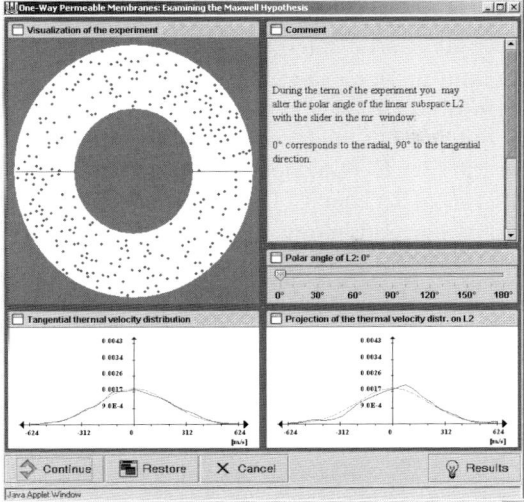

E 6.2. Projected velocity distributions

6.7 Conclusions

The angular velocity of the fluid increases, with the consequence that the temperature decreases.

The increase of $\omega(t)$ is limited, as a consequence of the law of conservation of energy. The theoretically maximal angular velocity is not reached within

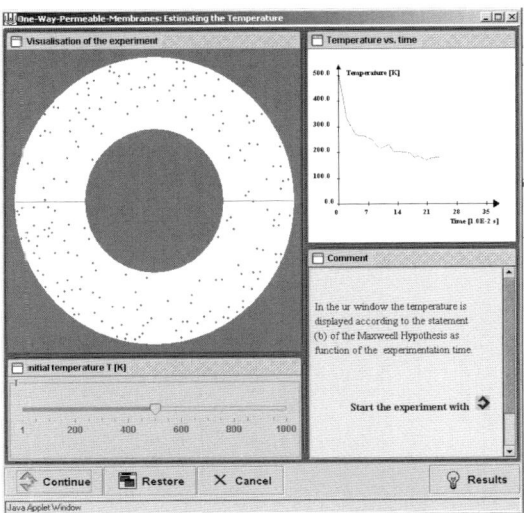

E 6.3. Temperature versus time

the experiments, nor does the temperature reach 0 [K]. As the angular velocity of the fluid increases, it can be shown experimentally that the molecules are concentrated more and more at the periphery of the ring-shaped container.

Of course, the design of the experiment defines a thermodynamic machine having only one pool of heat, which means a perpetuum mobile of the second kind, contradicting Kelvin's formulation of the second law of thermodynamics, cf. Reif (1965), Section 5.11. This unreality could also be confirmed by showing that the thermodynamical entropy of the fluid is decreasing as function of time.

7 (M, M)-Dynamics

7.1 A Generalization of the Mass Concept

For a Boltzmann system of moving molecules in $\mathbb{R}^d$, $d = 2, 3$, on which Newtonian dynamics is imposed, we have as the relation between a velocity vector v and the corresponding momentum vector u

$$(7.1.1) \qquad\qquad u = M \cdot v,$$

while the kinetic energy $E_{\text{kin}}(v)$ as function of the velocity is given by

$$(7.1.2) \qquad\qquad E_{\text{kin}}(v) = \frac{1}{2} v' M v,$$

the mass matrix M for the case $d = 2$ being given by

$$(7.1.3) \qquad\qquad M = \begin{pmatrix} m & 0 \\ 0 & m \end{pmatrix} = m \begin{pmatrix} 1 & 0 \\ 0 & 1 \end{pmatrix}.$$

Obviously we have

$$(7.1.4) \qquad\qquad v = M^{-1} u,$$

while the kinetic energy $E_{\text{kin}}(u)$ as function of the momentum is given by

$$(7.1.5) \qquad\qquad E_{\text{kin}}(u) = \frac{1}{2} u' M^{-1} u$$

with

$$(7.1.6) \qquad\qquad M^{-1} = \begin{pmatrix} 1/m & 0 \\ 0 & 1/m \end{pmatrix} = \frac{1}{m} \begin{pmatrix} 1 & 0 \\ 0 & 1 \end{pmatrix}.$$

Up to multiplication by the positive scalar $m > 0$ (mass of a molecule) the mass matrix is given in the Newtonian set-up as a unit matrix.

A generalization of this Newtonian concept allows us to examine which of the physical facts discussed hitherto, remain valid even under the generalized conditions, or which are only true in the special Newtonian case, respectively.

Thus it must be seen that M acts in a double sense. Firstly it determines the quadratic form describing the kinetic energy as function of the velocity and secondly, taken as a linear mapping, it determines the momentum vector when the velocity vector is given. So we will separate these two functions of the **mass matrix** M. On the one hand we introduce the **energy matrix** M as a positive definite and symmetric matrix. Representing the kinetic energy as in (7.1.2), a non-symmetric matrix M may always be substituted by a symmetric one. On the other hand we define the **momentum matrix** B as an invertible matrix B with inverse B^{-1}.

When a generalized dynamics is implemented on the computer, the conditions of conservation of momentum and energy still remain valid.

To differentiate from standard Newtonian dynamics, we will speak of (generalized Newtonian) (M, B)-dynamics, respectively of (generalized Newtonian) (M, M)-dynamics, if $M = B$ holds true. A system of moving molecules with (M, M)- and (M, B)-dynamics is called a (M, M)- and a (M, B)-system of moving molecules, respectively.

In the sequel our basis is primarily the momentum distribution. On the basis of Equation (7.1.2) it follows from

$$u \; = \; Bv$$

that the matrix defining the kinetic energy as a function of the momentum in the (M, B)-case is

(7.1.7) $$B_T^{-1} M B^{-1}$$

instead of M^{-1} in the (M, M)-case.

7.2 Kinetic Dynamics

Our theoretical basis of Kinetic Dynamics is as in 1.8, where we have here only one kind of molecule whose energy matrix M and momentum matrix B are as in Section 7.1.

The representation of positions $x^{(i)}$ and velocities $v^{(i)}$ of molecules corresponds to those in 1.8.

The time $\bar{t}$ of the next collision or reflection is again defined as in (1.8.2):

(7.2.1) $$\bar{t} := \min\left(\{t_i | i \in \mathbb{N}_N\} \cup \{t_{ij} | i, j \in \mathbb{N}_N, i < j\}\right).$$

This holds also true for the unit vector of the direction of the momentum exchange:

(7.2.2) $$e := \frac{x^{(i)}(\bar{t}) - x^{(j)}(\bar{t})}{|x^{(i)}(\bar{t}) - x^{(j)}(\bar{t})|}.$$

In order to determine the outcome of a collision, let $u^{(i)}$, $u^{(j)}$ and $\bar{u}^{(i)}$, $\bar{u}^{(j)}$ denote the momentum of the molecules i and j immediately before and immediately after the impact at time $\bar{t}$.

The conservation of momentum implies that

$$(7.2.3) \qquad \begin{aligned} \bar{u}^{(i)} &= u^{(i)} + \xi e \text{ and} \\ \bar{u}^{(j)} &= u^{(j)} - \xi e \end{aligned}$$

with ξ being determined by the condition of the conservation of energy

$$(7.2.4) \qquad E(\bar{u}^{(i)}) + E(\bar{u}^{(j)}) = E(u^{(i)}) + E(u^{(j)}),$$

where the energy is represented here as function of momentum. According to this we calculate

$$(7.2.5) \qquad \begin{aligned} 2\,E(\bar{u}^{(i)}) + 2\,E(\bar{u}^{(j)}) = \; & (u^{(i)} + \xi e)' B_T^{-1} M B^{-1}(u^{(i)} + \xi e) \\ & + (u^{(j)} - \xi e)' B_T^{-1} M B^{-1}(u^{(j)} - \xi e). \end{aligned}$$

Because of

$$(7.2.6) \qquad (u^{(k)} \pm \xi e)' B_T^{-1} = (B^{-1}(u^{(k)} \pm \xi e))', \qquad k = i, j,$$

the right hand term of (7.2.5) can be written as

$$(7.2.7) \qquad \begin{aligned} &\left(B^{-1}(u^{(i)} + \xi e)\right)' M B^{-1}(u^{(i)} + \xi e) \\ &+ \left(B^{-1}(u^{(j)} - \xi e)\right)' M B^{-1}(u^{(j)} - \xi e) \\ =\; & \left[(B^{-1}u^{(i)})' + \xi(B^{-1}e)'\right] M \left[B^{-1}u^{(i)} + \xi B^{-1}e\right] \\ &+ \left[(B^{-1}u^{(j)})' - \xi(B^{-1}e)'\right] M \left[B^{-1}u^{(i)} - \xi B^{-1}e\right] \\ =\; & \left[v^{(i)'} + (\xi B^{-1}e)'\right] M \left[v^{(i)} + \xi B^{-1}e\right] \\ &+ \left[v^{(j)'} - \xi(B^{-1}e)'\right] M \left[v^{(j)} - \xi B^{-1}e\right] \\ =\; & v^{(i)'} M v^{(i)} + \xi v^{(i)'} M B^{-1}e + \xi e' B_T^{-1} M v^{(i)} + \xi^2 e' B_T^{-1} M B^{-1}e \\ & + v^{(j)'} M v^{(j)} - \xi v^{(j)'} M B^{-1}e \\ & - \xi e' B_T^{-1} M v^{(j)} + \xi^2 e' B_T^{-1} M B^{-1}e. \end{aligned}$$

The right hand term of (7.2.7) equals $2\,E\bar{u}^{(i)} + 2\,E\bar{u}^{(j)}$, while $v^{(i)'} M v^{(i)}$ and $v^{(j)'} M v^{(j)}$ in the same term represent $2\,E(v^{(i)})$ and $2\,E(v^{(j)})$, respectively. By the law of the conservation of energy we obtain therefore

$$(7.2.8) \quad 2\xi^2 e' B_T^{-1} M B^{-1}e + \xi(v^{(i)} - v^{(j)})' M B^{-1}e + \xi e' B_T^{-1} M(v^{(i)} - v^{(j)}) = 0.$$

As the factor of ξ in the third summand of the left hand side of (7.2.8) is an element of $\mathbb{R}$, i.e.

$$e' B^{-1} M(v^{(i)} - v^{(j)}) \in \mathbb{R},$$

and as M is assumed to be symmetric, we recognize that the second and third summand of the left hand side of (7.2.8) are identical. Therefore dividing by $\xi \neq 0$ in (7.2.8) we obtain

$$(7.2.9) \qquad \xi = -\frac{(v^{(i)} - v^{(j)})'MB^{-1}e}{e'B_T^{-1}MB^{-1}e}.$$

Note that the conditions of momentum and energy conservation are not violated for the dynamics developed even if the momentum matrix B is non-symmetric.

For the special case of B being symmetric and equal to M we obtain

$$(7.2.10) \qquad \xi = -\frac{(v^{(i)} - v^{(j)})'e}{e'M^{-1}e}.$$

In order to determine the outcome of the reflection of the i-th molecule off the boundary, $u^{(i)}$ and $\bar{u}^{(i)}$ again denote the momentum before and after the reflection, while e_b denotes the unit vector orthogonal to the boundary at the reflection point, describing the exchange direction.

Based on the ansatz

$$(7.2.11) \qquad \bar{u}^{(i)} = u^{(i)} + \xi e_b$$

we obtain from the condition of energy conservation

$$(7.2.12) \qquad E(\bar{u}^{(i)}) = E(u^{(i)})$$

the following value for ξ:

$$\xi = -2\frac{v^{(i)'}MB^{-1}e_b}{e_b'B_T^{-1}MB^{-1}e_b}.$$

For the special case of B being symmetric and equal to M we have

$$(7.2.13) \qquad \xi = -2\frac{v^{(i)'}e_b}{e'M^{-1}e_b}.$$

7.3 Separating Dynamics

Computer experiments of colliding molecules on which Newtonian (M, B)-dynamics is imposed, show that molecules may 'stick' together after an impact or they may stick at the boundary after a contact with it.

Contrary to this – such phenomena could not be observed by us in experiments with either standard Newtonian or relativistic dynamics.

Let $v^{(i)}, v^{(j)}$ and $\bar{v}^{(i)}, \bar{v}^{(j)}$ be the velocities of two colliding molecules shortly before and shortly after their collision, and v and $\bar{v}$ the velocities of a molecule shortly before and shortly after its reflection off the boundary.

Definition 7.3.1

Let a certain dynamics be imposed on a system of molecules. Then the dynamics is called **separating**, if

(7.3.1)
$$\langle\, v^{(i)} - v^{(j)}, e\,\rangle \;<\; 0 \;\Rightarrow\; \langle\, \bar{v}^{(i)} - \bar{v}^{(j)}, e\,\rangle > 0 \text{ and}$$
$$\langle\, v, e_b\,\rangle \;<\; 0 \Rightarrow \langle\, \bar{v}, e_b\,\rangle \;>\; 0$$

holds true for all colliding pairs (i, j) of molecules and of all molecules hitting the boundary of the container, respectively.

If a dynamics is separating, the phenomenon of molecules sticking together is – for geometrical reasons – not possible.

Subject to (M, B)-dynamics for the case of **a collision** – with $v = B^{-1}u$ in mind – we obtain

(7.3.2) $\langle\, \bar{v}^{(i)} - \bar{v}^{(j)}, e\,\rangle \;=\; \langle\, v^{(i)} - v^{(j)}, e\,\rangle - 2\,\dfrac{\langle\, v^{(i)} - v^{(j)}, MB^{-1}e\,\rangle}{\langle\, e, b_T^{-1}MB^{-1}e\,\rangle}\cdot\langle\, e, B^{-1}e\,\rangle$

from (7.2.3) and (7.2.9). Similarly, for the case of a **reflection**,

(7.3.3) $\qquad\langle\, \bar{v}, e_b\,\rangle \;=\; \langle\, v, e_b\,\rangle \;-\; 2\,\dfrac{\langle\, v, MB^{-1}e\,\rangle}{\langle\, e, B_T^{-1}MB^{-1}e\,\rangle}\,\langle\, e, B^{-1}e\,\rangle.$

For (M, M)-dynamics we obtain for a **collision**

(7.3.4) $\qquad\qquad\langle\, \bar{v}^{(i)} - \bar{v}^{(j)}, e\,\rangle \;=\; -\,\langle\, v^{(i)} - v^{(j)}, e\,\rangle$

and, for the case of a **reflection**,

(7.3.5) $\qquad\qquad\qquad\langle\, \bar{v}, e\,\rangle \;=\; -\,\langle\, v, e\,\rangle.$

By (7.3.4) and (7.3.5), the condition (7.3.1) is fulfilled; this means that (M, M)-dynamics is separating.

In the sequel we treat (M, M)-dynamics, while the examination of (M, B)-dynamics is the subject of Chapter 8.

7.4 Entropic Distributions

According to 2.2 and 2.3, the notion of the temperature as a scalar quantity results from the coincidence of the ergodic and a specific, i.e. the accurate entropic momentum distribution which gives rise to study the family of the entropic momentum distributions. The approach based on velocity instead of momentum leads to corresponding results.

Let

(7.4.1) $\qquad(\varOmega, \mathcal{A}) := \left(\bigtimes_{j=1}^{N} \mathbb{U}^{(j)}, \bigotimes \mathcal{U}^{(j)}\right) \;=\; \left(\mathbb{R}^{dN}, \mathcal{B}^{dN}\right)$

be the measurable space of momentum, where in our experiments we set $d = 2$.

In order to study the momentum distribution in the case of (M, M)-dynamics, the measurable function $H: \Omega \to \mathbb{R}_+$ from Theorem 2.1.3 is defined as

$$(7.4.2) \quad \begin{cases} H: \mathbb{R}^{2N} \longrightarrow \mathbb{R}_+ & \text{with} \\ H(u) = \sum_{j=1}^{N} H_0(u^{(j)}), & u := (u^{(1)}, \dots, u^{(N)}) \in \mathbb{R}^{dN}, \end{cases}$$

i.e., H is additively representable, while the function H_0 is defined as

$$(7.4.3) \quad \begin{aligned} H_0: \mathbb{R}^2 &\longrightarrow \mathbb{R} \\ H_0(w) &:= \frac{1}{2} w' M^{-1} w, \quad w \in \mathbb{R}^d. \end{aligned}$$

$H_0(w)$ may be interpreted as the kinetic energy of a molecule; $H(u)$ that of N molecules.

Performing a standard integration we obtain for $\beta > 0$

$$(7.4.4) \quad Z_0(\beta) = \int_{\mathbb{R}^d} \exp(-\beta H_0(w)) \, d\lambda^d(w) = \left((|M|)(2\pi/\beta)^d \right)^{\frac{1}{2}},$$

$$(7.4.5) \quad Z_N(\beta) = \int_{\mathbb{R}^{dN}} \exp(-\beta H(u)) \, d\lambda^{dN}(u) = ((|M|)(2\pi/\beta)^d)^{N/2},$$

$|M|$ denoting the determinant of M. Comparing (7.4.4) and (7.4.5) we get

$$(7.4.6) \quad Z_N(\beta) = (Z_0(\beta))^N \quad (\beta > 0).$$

Analogously we obtain the expected kinetic energy of N molecules

$$(7.4.7) \quad \begin{aligned} \varepsilon(\beta) &= \frac{1}{Z_N(\beta)} \int_{\mathbb{R}^{dN}} H(u) \exp(-\beta H(u)) \, d\lambda^{dN}(u) \\ &= \frac{N}{Z(\beta)} \int_{\mathbb{R}^d} H_0(w) \exp(-\beta H_0(w)) 5 \, d\lambda^d(w) = \frac{Nd}{2\beta} < \infty. \end{aligned}$$

The density

$$(7.4.8) \quad f_\beta: \bigtimes_{j=1}^{N} U^{(j)} = \mathbb{R}^{dN} \longrightarrow \mathbb{R}_+$$

with

$$\begin{aligned}
f_\beta(u^{(1)}, \ldots, u^{(N)}) &= \left(\frac{1}{Z_N(\beta)} \exp(-\beta H(u^{(1)}, \ldots, u^{(N)})) \right) \\[2mm]
&= \frac{1}{Z_N(\beta)} \exp\left(-\frac{\beta}{2} \sum_{j=1}^{N} H_0(u^{(j)}) \right) \\[2mm]
&= \prod_{j=1}^{N} \frac{1}{Z_0(\beta)} \exp(-\beta H_0(u^{(j)})) \\[2mm]
&= \prod_{j=1}^{N} (|M|(2\pi/\beta)^d)^{-\frac{1}{2}} \exp\left(-\frac{\beta}{2} u^{(j)\prime} M^{-1} u^{(j)} \right)
\end{aligned}$$

(7.4.9)

determines a probability measure P_β on $\mathcal{B}^{dN}$, being the N-th power of a probability measure P_β^0 on $\mathcal{B}^d$:

(7.4.10) $$P_\beta = f_\beta \lambda^{dN} = (P_\beta^0)^N = N(0, M/\beta)^N .$$

A λ^d-density f_β^0 of P_β^0 is given by

(7.4.11) $$f_\beta^0(w) = (|M|(2\pi/\beta)^d)^{-\frac{1}{2}} \exp\left(-\frac{\beta}{2} w' M^{-1} w \right) .$$

The families $\{P_\beta | \beta > 0\}$ and $\{P_\beta^0 | \beta > 0\}$ of probability measures P_β and P_β^0 on $\mathcal{B}^{dN}$ and $\mathcal{B}^d$ are the families of the *entropic* **momentum** *distributions* of the whole system of N molecules and of just one molecule parametrized by β, respectively.

For P_β and P_β^0 we also write $P_{\mathrm{ent}}(\beta)$ and $P_{\mathrm{ent}}^0(\beta)$, respectively.

Defining

(7.4.12) $$\varepsilon_0(\beta) := \frac{1}{Z_0(\beta)} \int_{\mathbb{R}^d} H_0(w) \exp(-\beta H_0(w)) \, d\lambda^d(w)$$

as the expected kinetic energy of one molecule we obtain from (7.4.7)

(7.4.13) $$\varepsilon(\beta) = N \varepsilon_0(\beta).$$

According to Theorem 2.1.3, P_β is the optimizer of the Boltzmann–Gibbs entropy with respect to all λ^{dN} dominated probability measures Q on $\mathcal{B}^{dN}$ satisfying

(7.4.14) $$\int H \, dQ = \varepsilon(\beta) = N \varepsilon_0(\beta).$$

7.5 Temperature and Pressure

We have already proved that (M, M)-dynamics, as a straightforward generalization of standard Newtonian dynamics, is separating.

The questions with regard to further properties of the (M, M)-dynamics
are suggested by well-known experiences of the kinetic gas theory of Maxwell
and Boltzmann, such as the question about the temperature. Is the temperature still a scalar quantity or has a 'directional temperature' to be introduced?

Establishing a notion of pressure according to the one in Section 5.1, the
question now arises whether for a symmetric positive definite mass matrix M,
the pressure depends on the localization of the manometer within a container;
i.e. whether the pressure is an invariant.

Also, certainly of importance is the question whether the so-called equation of state

$$(7.5.1) \qquad\qquad p \; = \; k_\mathrm{B} \, \frac{N \cdot T}{V} \, ,$$

which relates pressure p and temperature T of N moving molecules within
a container of volume V, can be reformulated for the case of a (M, M)-system.

7.6 The Course of Experimentation

To enter into a systematic discussion of the questions formulated in 7.5, a first
requirement is a computer-experimental examination of the existence of an
ergodic distribution of velocities or momenta for (M, M)-dynamics, cf. E 7.1.

Contrary to the Maxwell Hypothesis, statement (a), cf. (1.1.11) or (1.5.7),
the ergodic distributions of velocities or momenta of a molecule are given as
bivariate normal distributions which in general are not rotational symmetric.

The technique of statistical analysis in experiment E 7.1 is essentially
the same as in experiment E 1.1, cf. 1.6 'Design of the Experiment' and
1.7 'Estimation Methods'. To this end, let L_φ again be the (1-dimensional)
linear subspace, with unit vector e_φ, of $\mathbb{V}$ (or $\mathbb{U}$) of velocities (or momenta)
of a molecule, while π_φ denotes the projection of $\mathbb{V}$ onto $L_\varphi \subset \mathbb{V}$, (or of $\mathbb{U}$
onto $L_\varphi \subset \mathbb{U}$), respectively.

The aim is to estimate the distribution of both the identically distributed
random variables

$$(7.6.1) \qquad\qquad V_\varphi^{(j)} \; := \; \pi_\varphi \circ V^{(j)} \, , \quad j = 1, \ldots, N \, ,$$

and the identically distributed random variables

$$(7.6.2) \qquad\qquad U_\varphi^{(j)} \; := \; \pi_\varphi \circ U^{(j)} \, , \quad j = 1, \ldots, N \, ,$$

i.e. to estimate the projections of the velocity and momentum distributions
of a molecule onto L_φ, respectively.

In experiment E 7.1 we determine – according to experiment E 1.1 – the kernel density estimates of these projections and compare them with a centered normal distribution whose variance is estimated according to (1.7.1).

The result of experiment E 7.1 is decisive for the further analysis: the realized velocity or momentum vectors form an ellipsoid in the velocity or momentum space of a molecule, respectively. The variances of the projections of the ergodic velocity or momentum distribution of a molecule onto L_φ depend on the polar angle φ of L_φ, where those projected distributions with the minimal or maximal variances can be related to the (presumed) main axis of the ellipsoid. This proves that the ergodic velocity or momentum distribution of a molecule are centered, bivariate normal distributions.

In experiment E 7.2 we confine ourselves to the momentum distributions of a molecule.

There we take as a basis a slightly different method for the estimation of the variance of the projected (ergodic) momentum distribution of a molecule onto the linear subspace L_φ.

Let

$$(7.6.3) \qquad \mathrm{Cov}(U^{(j)}) \;=\; E(U^{(j)}U^{(j)'})$$

be the covariance matrix of the identically distributed random momentum vectors $U^{(j)}$ of the j-th molecule, $j = 1, \ldots, N$.

Since

$$(7.6.4) \qquad U_\varphi^{(j)} \;=\; \pi_\varphi \circ U^{(j)} \;=\; \langle\, e_\varphi, U^{(j)} \,\rangle$$

the covariance $\mathrm{Var}(\pi_\varphi \circ U^{(j)})$ of the projection of the distribution of $U^{(j)}$ onto L_φ can be expressed as

$$(7.6.5) \quad \begin{aligned} \mathrm{Var}(\pi_\varphi \circ U^{(j)}) \;&=\; E\big((e_\varphi' U^{(j)})(U^{(j)'} e_\varphi)\big) \;=\; E(e_\varphi' U^{(j)} U^{(j)'} e_\varphi) \\ &=\; e_\varphi' E(U^{(j)} U^{(j)'}) e_\varphi \;=\; e_\varphi' \,\mathrm{Cov}(U^{(j)}) e_\varphi \,. \end{aligned}$$

As $\mathrm{Cov}(U^{(j)})$ is not known, we use a consistent estimator $\mathrm{cov}(U^{(j)})$ of $\mathrm{Cov}(U^{(j)})$ being given by

$$(7.6.6) \qquad \mathrm{cov}(U^{(j)}) \;=\; \begin{pmatrix} \mathrm{cov}_{11}(U^{(j)}), & \mathrm{cov}_{12}(U^{(j)}) \\ \mathrm{cov}_{12}(U^{(j)}), & \mathrm{cov}_{22}(U^{(j)}) \end{pmatrix}$$

with

$$\mathrm{cov}_{ii}(U^{(j)}) \;=\; \frac{1}{N} \sum_{j=1}^{N} U^{(j)^2} \,, \quad i = 1, 2 \,,$$

$$\mathrm{cov}_{12}(U^{(j)}) \;=\; \mathrm{cov}_{21}(U^{(j)}) \;=\; \frac{1}{N} \sum_{j=1}^{N} U_1^{(j)} U_2^{(j)} \,.$$

In experiment E 7.2 we examine whether the Entropy Principle (2.3.2), statement (a), still holds true. The latter requires the coincidence of the ergodic momentum distribution P_{erg}^0 (of a molecule) with an entropic momentum distribution $P_{\text{ent}}^0(\beta)$ (of a molecule) for some specific $\beta^* \in \mathbb{R}$. By (7.4.10) and (7.4.11) the entropic momentum distributions $P_\beta^0 = P_{\text{ent}}^0(\beta)$, $\beta > 0$, are centered, bivariate normal distributions, their covariance matrices being given by

$$(7.6.7) \qquad\qquad \text{Cov}\, P_{\text{ent}}^0 \;=\; M/\beta, \quad \beta > 0.$$

In order to validate the Entropy Principle we have therefore to examine whether there exists a $\beta^* \in \mathbb{R}$ such that

$$(7.6.8) \qquad\qquad \text{Cov}(U^{(j)}) \;=\; M/\beta^*.$$

In experiment E 7.2 we go a step further and compare the variances of the projections of the family of entropic distributions of a molecule onto $L_\varphi, 0° \leq \varphi \leq 360°$, with the corresponding projections of the ergodic distribution.

If the solution β_φ^* of

$$(7.6.9) \qquad\qquad \frac{1/\beta_\varphi^*(e_\varphi' M e_\varphi)}{e_\varphi' \, \text{cov}(U^{(j)}) e_\varphi} \;=\; 1$$

should be the same for all $\varphi, 0° \leq \varphi \leq 360°$, we would prove experimentally the validity of the Entropy Principle, part (a), cf. 2.3.2 Entropy Principle, for some β^*.

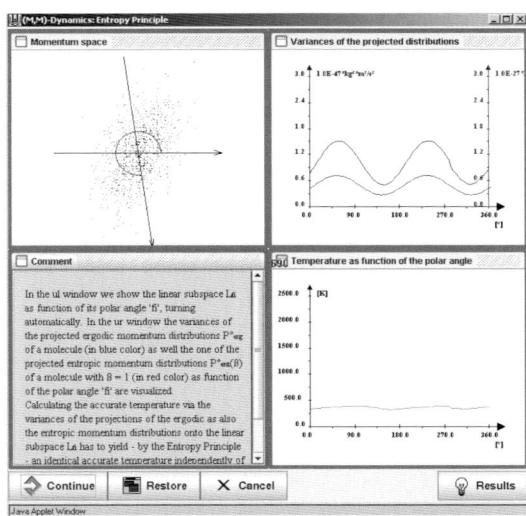

E 7.2. Entropy principle and temperature

Determining the temperature according to the Entropy Principle, statement (b), by

$$(7.6.10) \qquad\qquad T = 1/(k_B \cdot \beta)$$

for some $\beta > 0$ and using (7.6.9), we arrive at

$$(7.6.11) \qquad T^* := T^*_\varphi = \frac{1}{k_B} \frac{e'_\varphi \operatorname{cov}(U^{(j)}) e_\varphi}{e'_\varphi M e_\varphi}, \quad 0° \le \varphi \le 360°.$$

This is precisely the result of experiment E 7.2 showing not only the validity of the Entropy Principle but also the existence of a temperature as a scalar quantity even for the (M, M)-system of moving molecules.

Indeed, T^*_φ in (7.6.11) could be interpreted as the 'directional temperature' in the direction defined by the polar angle φ. In this sense (7.6.11) states that all directional temperatures T^*_φ are equal to the one temperature T^*.

During experiment E 7.3 it will be examined whether the pressure depends on the position of a manometer within a container, i. e. whether the pressure is an invariant.

The experiment displays the evolution of the pressure measurement of two differently positioned manometers, so that these measurements can be compared with the prediction of the pressure based on the equation of state, cf. (7.5.1).

As the measurements of the pressure coincide with the prediction based on the equation of state, it can be stated that the (M, M)-system of moving molecules shows the same phenomena as the Boltzmann system.

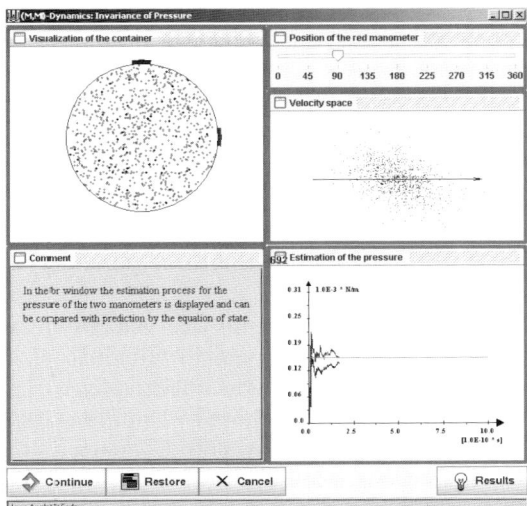

E 7.3. Invariance of pressure

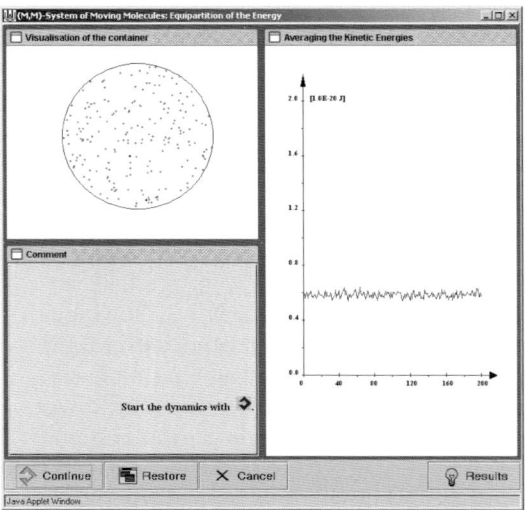

E 7.4. Equipartition of energy

Although the invariance of pressure and also its predictability by the equation of state can also be restated for relativistic dynamics, it can not be taken as a matter of course. In comparison to Chapter 7, the slightly modified dynamics of Chapters 8 and 9 show other results.

Experiment E 7.4 shows the equipartition of the energy with respect to the N molecules of the (M, M)-system, i. e. the expected energies of the molecules, as calculated by an averaging process of the realized energies, are identical for all molecules.

The result of experiment E 7.4 is confirmed for almost any of the considered dynamics; an expectation is the stochastically determined energy splitting of experiment E 12.4, when the parameter t defined there is not 0.5.

7.7 Conclusions

The system with (M, M)-dynamics is a natural generalization of the Boltzmann system of moving molecules.

The ergodic distributions (of momenta and velocities) are centered, normal distributions, not necessarily rotationally symmetric, cf. experiment E 7.1. That is, the ergodic distributions coincide statistically with centered normal distributions, their variances being estimated within the class of centered normal distributions based on data projected onto L_φ, i. e. with the corresponding accurate entropic distributions. The results of the graphical exploration were partially tested based on a χ^2-goodness-of-fit test.

Even so the temperature is – astonishingly enough – still a scalar quantity; i. e. experiment E 7.1 and E 7.2 prove the validity of the Entropy Principle for (M, M)-dynamics. Note that this result therefore represents an insight which is not already covered by experiences gained from the real world.

Experiments E 7.3 and E 7.4 confirm as expected the validity of the equation of state and the equipartition of energy, respectively.

8 (M, B)-Dynamics

8.1 Tunneling

We have already pointed out in Section 7.1 that the mass matrix M as defined in (7.1.3) acts in a double sense. At first determines the quadratic form describing the energy and secondly – taken as a linear mapping – it determines the momentum vector for a given velocity vector.

This prompted us to introduce (M, B)-dynamics with the energy matrix M (representing the energy with respect to velocity!) as a positive definite and symmetric matrix, as well as the momentum matrix B as an invertible matrix.

In Section 7.2 we developed the kinetic dynamics for the energy and momentum concept described by the matrices M and B with regard to a software implementation, cf. also (7.1.7).

In contrast to (M, M)-dynamics, (M, B)-dynamics is not separating – as we know from experiments. This means that two colliding molecules may 'stick' together or that a molecule hitting the boundary may remain attached to it. This prompted us to give up in certain cases the fundamental fact that two bodies may not penetrate each other, which in a computer experiment is reduced to an adjustment of the software of kinetic dynamics in comparison with that of (M, M)-dynamics.

In experiment E 8.0 we consider a system of moving molecules subject to (M, B)-dynamics, where two colliding molecules may penetrate each other after an impact instead of impeding each other, as also where molecules after an impact with the boundary may penetrate the boundary instead of being impeded by it. The latter is reminiscent of quantum tunneling.

Experiment E 8.0 shows that the molecules leave the container unless $B = \lambda M$ for some $\lambda \neq 0$. An intuitive rule is that the container empties either slowly or quickly according as B is or is not well approximated by λM for some $\lambda \neq 0$.

In the further experiments E 8.1–E 8.4, (M, B)-dynamics will be modified insofar as two colliding molecules may indeed penetrate each other but the penetration of the boundary by molecules is no longer possible. This can (only) be achieved, when the hits of a molecule off the boundary are treated according to (M, M)-dynamics. This modification still guarantees energy conservation. We speak here of modified (M, B)-dynamics.

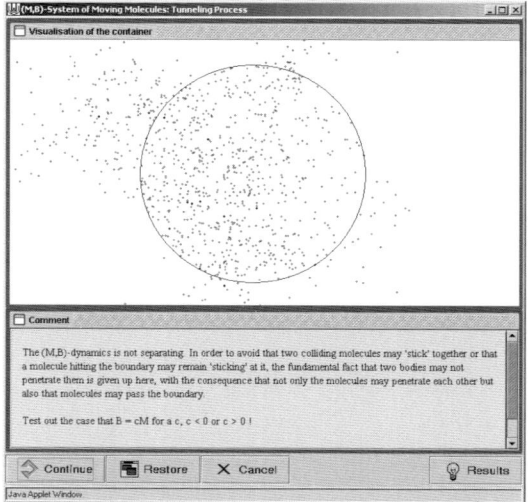

E 8.0. Tunneling

8.2 The Course of Experimentation

In experiment E 8.1 the existence of an ergodic momentum distribution will be examined. The design of the experiment corresponds to that of experiment E 7.1, and this also holds true for the result. The ergodic momentum distribution is a bivariate, centered normal distribution even in the present case.

The family of entropic distributions is – by definition – the same as in Chapter 7, so that experiment E 8.2, which follows the line of experiment E 7.2, shows the validity of the Entropy Principle which ensures the existence of a temperature as a scalar quantity.

The difference between experiment E 7.2 and experiment E 8.2 is on the one hand the dynamics imposed on the molecules. Experiment E 7.2 is based on (M, M)-dynamics while, in the present experiment, the modified (M, B)-dynamics is imposed. On the other hand, the family of entropic distributions is for the (M, B)-case, i.e. for the present experiment, different from that of the (M, M)-case, i.e. as in experiment E 7.2.

Since

$$u = Bv,$$

we find from (7.1.2) that the matrix defining the kinetic energy as function of momentum in the (M, B)-case is

$$B_T^{-1} M B^{-1}$$

while in the (M, M)-case this matrix is given by M^{-1}.

This means that the Hamiltonian, as it enters into the densities of the family of entropic distributions, is in the (M, B)-case different from the one in the (M, M)-case.

In experiment E 8.3 the invariance of the pressure is examined and compared with the prognosis based on the equation of state.

Experiment E 8.4 examines the equipartition of energy with respect to the molecules.

8.3 Conclusions

Experiment E 8.0 shows a tunneling effect; i. e. the molecules disappear from the container. Note that two colliding molecules in experiment E 8.0 may penetrate each other after an impact in order to not impede each other, and that a molecule after a contact with the boundary may pass through the boundary instead of being impeded by it.

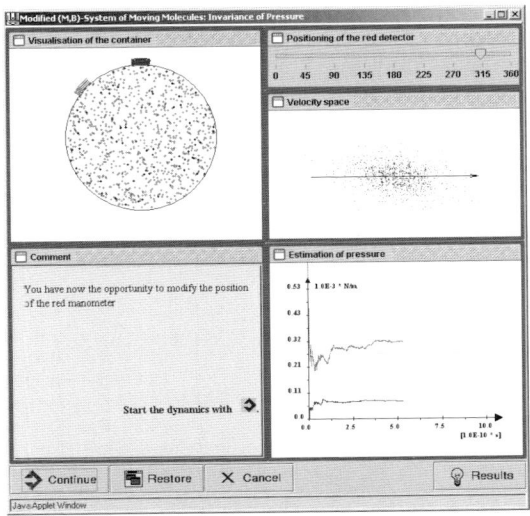

E 8.3. (Non)-Invariance of pressure

Indeed, it is not unimportant which energy and momentum functions are combined, but of course the question whether the Entropy Principle remains valid for (M, B)-dynamics is still open.

The modified (M, B)-dynamics also allows the penetration of two colliding molecules after an impact in order not to impede each other, but molecules are reflected at the boundary and cannot pass through it. The reflections of the molecules off the boundary can be established applying (M, M)-dynamics in these cases.

Experiments E 8.1 and E 8.2 with the modified (M, B)-dynamics show the same results as experiments E 7.1 and E 7.2, respectively; i. e. the substitution of the mass matrix M, taken as momentum matrix, by an invertible matrix B shows no consequences with respect to the existence and type of the ergodic distribution and the validity of the Entropy Principle.

As a main result we have that pressure is no longer invariant within the container; i. e. pressure depends on the localization of the manometer, so that the defining premise for the prediction of pressure by the equation of state is not given; cf. experiment E 7.3.

If the energy matrix M equals the momentum matrix B, then the experiment shows that pressure is an invariant and may be predicted by the equation of state according to experiment E 7.1.

If one sets $B = \lambda M$, for $\lambda \neq 0$, the pressure remains invariant; but the prognosis by the equation of state has to be multiplied now by λ in order to approximate the estimated pressure.

As the equation of state relates pressure (based on momentum) with temperature (related to energy), it is also theoretically clear that the validity of the equation of state can only be given when momentum and energy are based on the same mass matrix.

Experiment E 8.4 confirms the equipartition of energy with respect to the molecules as already experiment E 7.4 did for the (M, M)-case.

9 Acausal (M, M)-Dynamics

9.1 Non-Impact Related Energy Splitting

In the Boltzmann system of moving molecules based on Newtonian dynamics or on its generalizations, that is, (M, M)- and (M, B)-systems, the splitting of the sum of the kinetic energies of two colliding molecules is determined by the momentum exchange according to the law of momentum conservation.

The partners, i. e. the molecules involved, the time points, and the places of the energy splitting are determined by the impacts of the molecules, which initialize the energy splittings.

The link between the energy splitting of the molecules and the collision of two molecules is broken now. Thus energy splitting is no longer located at the place of the collision, while the laws of momentum and energy conservation – being a consequence of the Newtonian axioms – still remain valid. This especially means that the causality is given up – without a consequence for the statistical distributional aspects of the considered system, that is, for the Maxwell Hypothesis, part (a), as our experiments show. Indeed, the present investigations were motivated by a quantum problem.

A pair of molecules selected for energy splitting may – independently of their positions – split the sum of their kinetic energies determined by the laws of energy and momentum conservation, cf. (7.2.3), (7.2.4).

The selection of the molecules to split the energy is established by a random experiment based on the (discrete) uniform distribution on $\mathbb{N}_N \times \mathbb{N}_N$, where N denotes the number of molecules of the system. This indeed corresponds to the result of experiment E 3.6.

The time points of energy splitting – i. e. the inter-event times – are also outcomes of a random experiment based on the exponential distribution $\text{Exp}(\gamma)$ for some parameter γ, cf. experiment E 3.1. Along with this we still have the reflections of molecules off the boundary according to (M, M)-dynamics, cf. Section 7.2 'Kinetic Dynamics'.

We are speaking now of acausal (M, M)-dynamics.

9.2 The Course of Experimentation

The design of the experiments and the course, as well as the aims of the experimentation, are the same as in Chapter 7.

Experiment E 9.1 examines both the existence and the type of the ergodic distributions of momenta and velocity, while experiment E 9.2 checks the validity of the Entropy Principle.

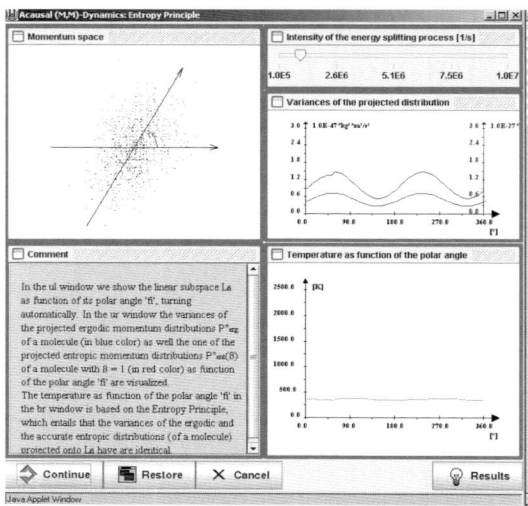

E 9.2. Entropy Principle (and temperature)

In all experiments, including E 9.3 (Invariance of Pressure) and E 9.4 (Equipartition of Energy), the experimenter has the possibility of fixing the collision rate, i.e. the parameter γ of the distribution $\mathrm{Exp}(\gamma)$ of the inter-impact times within a certain domain.

A further question to be treated within the framework of acausal (M, M)-dynamics, cf. experiment E 9.1, is whether the direction of momentum exchange, as defined in (7.2.2) by the unit vector e, can be altered; especially of course, we are interested in the consequences of such alterations on the empiric momentum and velocity distributions.

From the viewpoint of software techniques, this means substituting the unit vector e by a unit vector e' representing an alternative direction of momentum exchange, such that e and e' define a predetermined deviation angle. The dynamics thus modified is called AMED-dynamics: Alternative Momentum Exchange Direction.

Experiment E 9.5 follows the line of experiment E 9.1; the aim of experiment E 9.5 is to examine the ergodicity and shape of the empiric momentum and velocity distributions.

In experiment E 9.6 we go a step further. The polar angles φ of the artificial momentum exchange directions e' are outcomes of a random experiment for every single energy splitting of two molecules – based on the concept of momentum conservation ; the polar angles φ of e' are distributed uniformly on $[0°, 360°]$. We therefore speak of UDMED-dynamics (Uniformly Distributed Momentum Exchange Directions).

Experiment E 9.6 also follows the line of experiment E 9.1; the aim of E 9.6 is again to examine the ergodicity and shape of the empiric momentum and velocity distributions.

9.3 Conclusions

The results of experiments E 9.1, E 9.2 and E 9.4 correspond to those of E 7.1, E 7.2 and E 7.4, respectively.

No influence of the collision rate γ on the outcomes of the experiments could be observed – at least not within the domain implemented in the experiments.

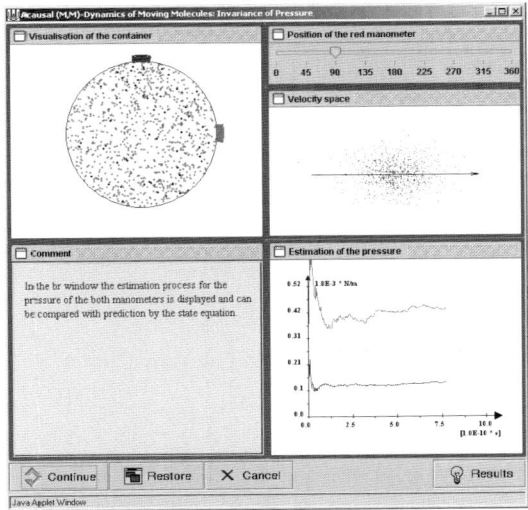

E 9.3. (Non-)Invariance of pressure

The invariance of pressure is only realized when the mass matrix M is a multiple of the unit matrix; in other cases the defining premises for the prognosis of pressure by the equation of state are not given.

If the manometers are positioned diametrically opposite to each other, they show the same (estimated) pressure – a phenomenon we cannot explain.

Experiment E 9.4 confirms the equipartition of energy with respect to the molecules.

The AMED- and UDMED-dynamics show in experiments E 9.5 and E 9.6, respectively, the same distributional results as experiment E 9.1; i. e. neither the alteration of the direction of momentum exchange nor its stochastification show any influence on the ergodic momentum distribution. Experiment E 9.6 must be seen in conjunction with experiment E 11.1.

We make a remark to the configuration space for acausal (M, M)-dynamics, this space is the container within which the molecules move. In contrast to (M, M)-dynamics, where the partner molecules of an energy splitting or the inter-event times are determined within the configuration space, this is no longer the case for acausal (M, M)-dynamics, where the partner molecules or the inter-event times are outcomes of, say, abstract, random experiments. In this respect the configuration space is no longer used.

10 Relativistic Dynamics

10.1 Relativistic Momentum and Energy

We again consider N molecules moving within a container B on which the relativistic dynamics is imposed. Fundamental for the considerations of Albert Einstein (14 March 1879–18 April 1955) was the condition

$$(10.1.1) \qquad v^{(j)}(t)| \ < \ c \,, \quad t \in \mathbb{R}_+ \,, \quad j = 1, \ldots, N \,,$$

for micro-constituents having a rest mass $m > 0$, where $c = 3 \times 10^8 [m/s]$ denotes the velocity of light.

In accordance with (10.1.1), the relativistic momentum of the j-th molecule is given by

$$(10.1.2) \qquad u^{(j)} := \frac{m \, v^{(j)}(t)}{\sqrt{1 - \frac{|v^{(j)}(t)|^2}{c^2}}} \,,$$

while the relativistic energy of a molecule is given by

$$(10.1.3) \qquad E^{(j)} := mc^2 \sqrt{1 + \frac{\langle u^{(j)}(t), u^{(j)}(t) \rangle}{m^2 c^2}} \,.$$

The number mc^2 is called the rest energy of a molecule and

$$(10.1.4) \qquad E_k^{(j)}(t) := E^{(j)}(t) - mc^2$$

the kinetic energy of the j-th molecule at time t.

The total energy of the gas with N micro-constituents, i. e. the Hamiltonian, is given by

$$H \colon \mathbb{R}^{2N} \ \to \ \mathbb{R}_+$$

$$(10.1.5) \qquad H(u) := \sum_{j=1}^{N} mc^2 \cdot \sqrt{1 + \frac{\langle u^{(j)}, u^{(j)} \rangle}{m^2 c^2}} \,, \quad u = (u^{(1)}, \ldots, u^{(N)}) \in \mathbb{R}^{2N} \,,$$

for molecules being given as non-overlapping discs with radius $r > 0$ and centers $x^{(j)}$:

$$(10.1.6) \qquad\qquad |x^{(i)} - x^{(j)}| \geq 2r.$$

H expresses the total kinetic energy of the system of molecules perceived by an observer for whom the container B does not move. Note that the relativistic Hamiltonian is additively representable.

More precisely one should speak of H as a quasi-relativistic Hamiltonian since the relativistic contraction phenomenon is neglected in its definition; this phenomenon entails the elliptical appearance of micro-constituents having here a disc-shaped eigengestalt. As we shall see in Section 10.2, neglecting the contraction phenomenon is a model simplification allowing an efficient implementation of relativistic dynamics to be imposed on the 2-dimensional system of moving molecules considered here.

10.2 Relativistic Kinetic Dynamics

Let us consider a fluid consisting of N identical discs of eigenradius $r > 0$ and rest mass $m > 0$ within a container B (cf. Section 10.1). Let us suppose that a micro-state of the gas at the time $t_0 \geq 0$, given by

$$\left(x(t_0); v(t_0) \right) = \left(x^{(1)}(t_0), \ldots, x^{(N)}(t_0); v^{(1)}(t_0), \ldots, v^{(N)}(t_0) \right),$$

where

$$|v^{(j)}(t_0)| < c, \qquad j = 1, \ldots, N,$$

is represented by a standard data structure. Let the positions $x^{(1)}(t), \ldots,$ $x^{(N)}(t)$ of the discs evolve according to

$$x^{(j)}(t) = x^{(j)}(t_0) + t\, v^{(j)}(t_0)$$

until the next collision occurs; during this time interval the velocities remain constant.

A collision can be a disc–disc or a disc–container collision; the determination of the next actual collision is analogous to Section 1.8 'Kinetic Dynamics'.

The outcome of a collision between molecules i and j at a time $\bar{t}$ can be computed by inserting the right hand terms of (10.2.1) (conservation of momentum)

$$(10.2.1) \qquad \begin{cases} \bar{u}^{(i)} = u^{(i)}(\bar{t}) + \xi e \\ \bar{u}^{(j)} = u^{(j)}(\bar{t}) - \xi e \end{cases}$$

into the energy equation

$$(10.2.2) \quad \sum_{k \in \{i,j\}} mc^2 \sqrt{1 + \frac{\langle \bar{u}^{(k)}, \bar{u}^{(k)} \rangle}{m^2 c^2}} = \sum_{k \in \{i,j\}} mc^2 \sqrt{1 + \frac{\langle u^{(k)}(\bar{t}), u^{(t)}(\bar{t}) \rangle}{m^2 c^2}}$$

where $u^{(k)}(\bar{t})$ denotes the momentum of the molecule k before the momentum exchange and $\bar{u}^{(t)}$ its momentum after the exchange, $k \in \{i, j\}$, and e denotes the unit vector describing the direction of momentum exchange, cf. (1.8.3). The parameter ξ can be determined by solving the Equation (10.2.2) which leads to a quadratic equation.

The outcome of a reflection is computed as in Section 1.8 for the Newtonian case. This treatment of the collisions and reflections guarantees the conservation of energy and momentum in the course of the computational process. According to our experimental experience, relativistic dynamics is separating.

10.3 Relativistic Entropic Distributions

Let $(\mathbb{U}^{(j)}, \mathcal{U}^{(j)}) := (\mathbb{R}^2, \mathcal{B}^2)$ be the momentum space of the j-th micro constituent, $j = 1, \ldots, N$. Analogously to Section 2.2, the measurable space of momentum of N molecules $(\Omega, \mathcal{A})$ is specified as

$$(\Omega, \mathcal{A}) := \left(\bigtimes_{j=1}^{N} \mathbb{U}^{(j)}, \bigotimes_{j=1}^{N} \mathcal{U}^{(j)} \right) = (\mathbb{R}^{2N}, \mathcal{B}^{2N}).$$

Now we observe that the (relativistic) Hamiltonian $H \colon \Omega \to \mathbb{R}_+$ (cf. 10.1.5) has the additive representation

$$(10.3.1) \qquad H(u) = \sum_{j=1}^{N} H_0(u^{(j)}), \qquad u = (u^{(1)}, \ldots, u^{(N)}) \in \mathbb{R}^{2N},$$

in which the function $H_0 \colon \mathbb{R}^2 \to \mathbb{R}_+$ is defined by

$$(10.3.2) \qquad H_0(w) := mc^2 \sqrt{1 + \frac{\langle w, w \rangle}{m^2 c^2}}, \qquad w \in \mathbb{R}^2.$$

A value $H_0(w)$ can be interpreted as the relativistic energy of a molecule having the momentum $w \in \mathbb{R}^2$, cf. (10.1.3). Performing a standard integration we obtain for $\beta > 0$

$$Z(\beta) := \int_{\mathbb{R}^2} \exp(-\beta H_0(w)) \, d\lambda^2(w)$$

$$(10.3.3)$$

$$= 2\pi m^2 c^2 \left(\frac{1}{mc^2 \beta} + \frac{1}{(mc^2 \beta)^2} \right) e^{-mc^2 \beta}.$$

By (10.3.1) we have

$$Z_N(\beta) := \int\limits_{\mathbb{R}^{2N}} \exp(-\beta H(u)) \, d\lambda^{2N}(u)$$

$$= \int\limits_{\mathbb{R}^{2N}} \exp\left(-\beta \sum_{j=1}^{N} H_0(u^{(j)})\right) d\lambda^{2N}(u^{(1)}, \ldots, u^{(N)})$$

$$= \int\limits_{\mathbb{R}^{2N}} \prod_{j=1}^{N} \exp(-\beta H_0(u^{(j)})) \, d\lambda^{2N}(u^{(1)}, \ldots, u^{(N)})$$

$$= \left(\int\limits_{\mathbb{R}^2} \exp(-\beta H_0(w)) \, d\lambda^2(w) \right)^N$$

for $\beta > 0$, and therefore

(10.3.4) $$Z_N(\beta) = (Z_1(\beta))^N$$

by (10.3.3). Analogously we obtain the expected energy of a molecule

(10.3.5)
$$\varepsilon_0(\beta) := \frac{1}{Z(\beta)} \cdot \int\limits_{\mathbb{R}^2} H_0(w) \exp(-\beta H_0(w)) \, d\lambda^2(w)$$

$$= mc^2 \cdot \frac{(mc^2\beta)^2 + 2mc^2\beta + 2}{(mc^2\beta)^2 + mc^2\beta}$$

for $\beta > 0$, and therefore the expected total energy of the system is given by

(10.3.6)
$$\varepsilon(\beta) := \frac{1}{Z_N(\beta)} \int\limits_{\mathbb{R}^{2N}} H(u) \exp(-\beta H(u)) \, d\lambda^{2N}(u)$$

$$= \frac{N}{Z_1(\beta)} \cdot \int\limits_{\mathbb{R}^2} H_0(w) \exp(-\beta H_0(w)) \, d\lambda^2(w) = N\varepsilon_0(\beta)$$

for $\beta > 0$.

The density

(10.3.7) $$f_\beta : \bigtimes_{j=1}^{N} U^{(j)} = \mathbb{R}^{2N} \longrightarrow \mathbb{R}_+$$

with

(10.3.8)
$$f_\beta(u^{(1)}, \ldots, u^{(N)}) = \frac{1}{Z_N(\beta)} \exp(-\beta H(u^{(1)}, \ldots, u^{(N)}))$$

$$= \frac{1}{Z_N(\beta)} \exp\left(-\beta \sum_{j=1}^{N} H_0(u^{(j)})\right)$$

$$= \prod_{j=1}^{N} \frac{1}{Z(\beta)} \exp(-\beta H_0(u^{(j)}))$$

determines a probability measure P_β on $\mathcal{B}^{2N}$ being the N-th power of a probability measure P_β^0:

$$(10.3.9) \qquad P_\beta := f_\beta \; \lambda^{2N} \; = \; (P_\beta^0)^N \,.$$

A λ^2-density f_β^0 of P_β^0 is given by

$$(10.3.10) \qquad f_\beta^0(w) \; = \; \frac{1}{Z(\beta)} \exp(-\beta H_0(w)) \,.$$

The family $\{P_\beta | \beta > 0\}$ and $\{P_\beta^0 | \beta > 0\}$ of probability measures on $\mathcal{B}^{2N}$ and $\mathcal{B}^2$ is the family of the relativistic entropic distributions for the system of N moving molecules and of just one molecule parametrized by β, respectively.

For P_β and P_β^0 we also write $P_{\text{ent}}(\beta)$ and $P_{\text{ent}}^0(\beta)$, respectively.

Note that $P_{\text{ent}}(\beta)$ is the entropy optimizer (2.1.4) for the relativistic Hamiltonian H introduced in (10.1.5); cf. Theorem 2.1.3.

The function $h_\beta^0 \colon \mathbb{R}_+ \longrightarrow \mathbb{R}_+$,

$$(10.3.11) \quad h_\beta^0(\rho) \; = \; \frac{\exp(mc^2\beta)}{m^2 c^2 \left(\frac{1}{mc^2\beta} + \frac{1}{(mc^2\beta)^2} \right)} \cdot \rho \exp\left(-mc^2\beta \sqrt{1 + \frac{\rho^2}{m^2 c^2}} \right)$$

represents a density of the momentum norm of a molecule.

10.4 The Course of Experimentation

As a first step, in experiment E 10.1 we prove computer-experimentally that a relativistic ergodic momentum distribution P_{erg}^0 of a molecule does exist, which of course is no longer a normal distribution. To this end, the relativistic dynamics was already developed in Section 10.2 for software implementation.

To avoid the technicality of numerical integration, we estimate in experiment E 10.1 not the (rotationally symmetric) ergodic momentum distribution P_{erg}^0 of a molecule, but instead its image under the Euclidean norm.

Applying the kernel density estimation to the realizations $|u^{(1)}(t)|, \dots,$ $|u^{(N)}(t)|$ of the independently, identically distributed momenta $U^{(1)}(t), \dots,$ $U^{(N)}(t)$ of the N molecules at times $t_1 < t_2 < \dots$, we get a sequence of density estimates of the distribution of the momentum norm of a molecule, which is displayed in experiment E 10.1. The density estimates stabilize in the course of the computational process, which provides statistical evidence for the existence of an ergodic distribution P_{erg}^0 of a molecule.

The Entropy Principle 2.3.2, statement (a), postulates the validity of

$$(10.4.1) \qquad P_{\text{erg}}^0 \; = \; P_{\text{ent}}^0(\beta)$$

for some $\beta^* > 0$, its examination being the subject of experiment E 10.2.

According to the Entropy Principle, statement (b), we get for the accurate temperature T^* of the relativistic system of moving molecules

(10.4.2) $$T^* := 1 \Big/ (k_B \cdot \beta^*) .$$

In experiment E 10.2 the estimated density of the momentum norm of the ergodic distribution P^0_{erg} of a molecule is compared with the densities, cf. (10.3.11), of the momentum norms of the entropic distributions $P^0_{\mathrm{ent}}(\beta)$ parametrized by β, in order to select a β^* for which (10.4.1) holds true.

Let $\hat{g}_N(t)$ be the kernel density estimate of the distribution of the momentum norm of a molecule for $t \gg 0$. Then the selected (estimated) value $\hat{\beta}$ of β^* is implicitly determined by the minimizer of the so-called variational distance

(10.4.3) $$\int |\hat{g}_N(t) - h^0_{\hat{\beta}}| \, \mathrm{d}\lambda^1 = \min_{\beta>0} \int |\hat{g}_N(t) - h^0_\beta| \, \mathrm{d}\lambda^1 ,$$

within the class of $(h^0_\beta)_{\beta>0}$ of the densities of the norms of the entropic distributions $P^0_{\mathrm{ent}}(\beta)$, $\beta > 0$.

The corresponding estimate $\hat{T}$ – according to the Entropy Principle 2.3.2, statement (b) – of temperature T^* is given by

(10.4.4) $$\hat{T} := 1/(k_B \cdot \hat{\beta}) .$$

Experiment E 10.2 indeed confirms the Entropy Principle, statement (a), based on the relativistic energy of a molecule, cf. (10.1.3), and justifies (10.4.4) to be the estimate of the temperature T^*.

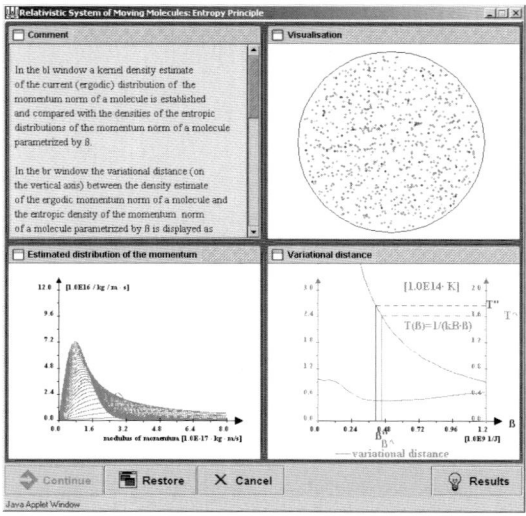

E 10.2. Entropy Principle

Also a second estimate presupposes the validity of the Entropy Principle, statement (a), confirmed by experiment E 10.2. By Formula (10.3.8) it is clear that the strong law of large numbers is valid for the sequence of averages

$$(10.4.5) \qquad \left(\frac{1}{N} \sum_{j=1}^{N} H_0(u^{(j)}(t)) \right)_{N \in \mathbb{N}}$$

approximating the expectation

$$(10.4.6) \qquad \int H_0 \, \mathrm{d}P_{\mathrm{erg}}^0 \;=\; \int H_0 \, \mathrm{d}P_{\mathrm{ent}}^0$$
$$= mc^2 \frac{(mc^2\beta^*)^2 + 2mc^2\beta^* + 2}{(mc^2\beta^*)^2 + mc^2\beta^*} \,.$$

The application of the strong law of large numbers entails now the consistency of (10.4.5) as an estimator of (10.4.6).

A consistent estimate β'' of β^* is given implicitly by

$$(10.4.7) \qquad \frac{1}{N} \sum_{j=1}^{N} H_0(u^{(j)}) \;=\; mc^2 \frac{(mc^2\beta'')^2 + mc^2\beta'' + 2}{(mc^2\beta'')^2 + mc^2\beta''} \,,$$

because β'' is a solution of the quadratic Equation (10.4.7); note that this solution depends continuously on the average (10.4.5). An estimate T'' of temperature T^* is given by

$$(10.4.8) \qquad T'' \;:=\; 1/(k_{\mathrm{B}} \cdot \beta'') \,.$$

In the further experiments of Chapter 10 the invariance of pressure is examined, cf. experiment E 10.3, while in E 10.4 the equipartition of energy with respect to the N molecules of the relativistic system is addressed.

10.5 Conclusions

It is certainly important to note that relativistic dynamics is – according to our collision experiments – a separating one.

In experiments E 10.1 and E 10.2, the Entropy Principle is proved experimentally for energy and momentum functions which are neither quadratic nor linear, respectively.

The empiric (i. e. ergodic) distribution of the relativistic momentum norm estimated in experiment E 10.1 coincides statistically with the accurate entropic distribution of the relativistic momentum norm. This result agrees with the outcome of a χ^2-goodness-of-fit test performed by us at a level of significance of 0.05.

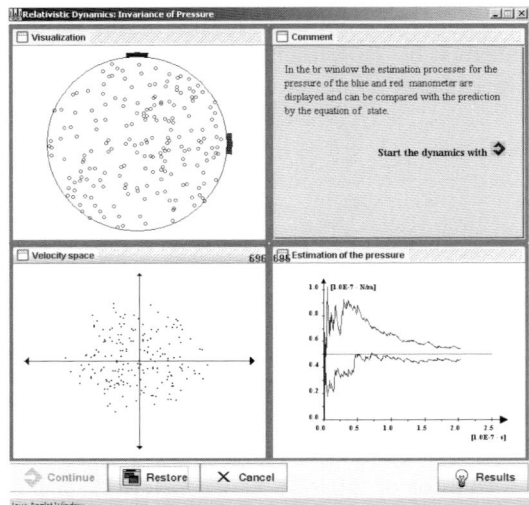

E 10.3. Invariance of pressure

Experiment E 10.2 yields two estimates of temperature based on different concepts and shows explicitly the validity of the Entropy Principle.

Experiments E 10.3 and E 10.4 show the same effects as experiments E 7.3 and E 7.4, respectively.

We note that differentiation of the relativistic free energy (not introduced here) with respect to the volume of the container yields the same result for the pressure term as in the non-relativistic case, which might be taken as an indication of the validity of the relativistic equation of state, which by definition requires the invariance of pressure.

11 Discrete Dynamics

11.1 A Discrete Momentum Space

Chapter 10 demonstrates the validity of the Entropy Principle even for momentum and energy functions which are not linear and not quadratic, respectively, while the experience from Chapter 9 indicates that we need not insist on the causality of the energy splitting process.

New here in Chapter 11 is the fact that the momentum space of the j-th molecule

$$(11.1.1) \qquad \mathbb{U}^{(j)} := \{\Delta u(\xi_1, \xi_2) \,|\, \xi_1, \xi_2 \in \mathbb{Z}\} \subset \mathbb{R}^2$$

for some $\Delta u > 0$, is discrete, i.e. it is a lattice, $j = 1, \dots, N$, with the consequence that, when applying Theorem 2.1.3, the dominating measure μ has to be discrete. In the sequel we substitute measure μ by the counting measure ζ_0 on $\mathbb{U}^{(j)}$.

Let the momentum vector of the j-th molecule be given by

$$u^{(j)} = (u_1^{(j)}, u_2^{(j)}) \in \mathbb{U}^{(j)}, \qquad j = 1, \dots, N.$$

Then the Hamiltonian H_0 of the j-th molecule is defined by

$$(11.1.2) \qquad H_0(u^{(j)}) := v_0(\kappa_1 \cdot |u_1^{(j)}| + \kappa_2 \cdot |u_2^{(j)}|), \qquad j = 1, \dots, N,$$

for $\kappa_1, \kappa_2 \in \mathbb{N}$ and $v_0 > 0$. The Hamiltonian H of N molecules is given by

$$H(u) = \sum_{j=1}^{N} H_0(u^{(j)}), \qquad u = (u^{(1)}, \dots, u^{(N)}).$$

By the definition of the Hamiltonian H_0 in (11.1.2), the entropy maximizing distribution of a momentum component is calculated based on Theorem 2.1.3. It reveals itself – cf. Section 11.3 – as a double geometric distribution on a lattice embedded in $\mathbb{R}$; its image under $u_i^{(j)} \mapsto v_0 \kappa_i |u_i^{(j)}|$, $i = 1, 2$, cf. (11.1.2), coincides with the equilibrium probabilities of the excited energy levels of the harmonic oscillator, cf. Reif (1965), Section 7.6.

11.2 Kinetic Discrete Dynamics

We consider a system of moving molecules confined to a rectangular container B on which a discrete dynamics is imposed. To introduce this dynamics we start with a momentum space according to (11.1.1).

The velocity space $\mathbb{V}^{(j)}$ of the j-th molecule is given as the image of $\mathbb{U}^{(j)}$ under the linear mapping induced by the matrix M^{-1} with M being given by

$$M := \begin{pmatrix} m & 0 \\ 0 & m \end{pmatrix}$$

for some $m > 0$, $j = 1, \ldots, N$.

A pair of molecules selected for energy splitting is determined by a random experiment based – according to the result of experiment E 3.6 – on the uniform distribution on $\mathbb{N}_N \times \mathbb{N}_N$, where N denotes the number of molecules of the system. The time points of energy splitting of molecules are outcomes of random experiments based on the exponential distribution $\text{Exp}(\gamma)$; the time points of energy splitting are independent of the positions of the molecules. Let $x^{(i)}, x^{(j)} \in B$ and $u^{(i)}, u^{(j)} \in \mathbb{U}^{(i)} = \mathbb{U}^{(j)}$, $i, j \in \mathbb{N}_N$, be the positions and momenta of the molecules i and j selected for the energy splitting, respectively, $i, j \in \mathbb{N}_N$.

Then the unit vector

$$(11.2.1) \qquad e := \frac{x^{(i)} - x^{(j)}}{|x^{(i)} - x^{(j)}|}$$

describes the direction of classical momentum exchange (cf. Section 1.8), which need not be compatible with the lattice structure of $\mathbb{U}^{(i)} = \mathbb{U}^{(j)}$.

Let $\bar{u}^{(i)}$, $\bar{u}^{(j)}$ denote the momenta of the molecules after the energy splitting.

The energy splitting is determined by the ansatz

$$(11.2.2) \qquad \begin{aligned} \bar{u}^{(i)} &= u^{(i)} + \bar{\Delta} \\ \bar{u}^{(j)} &= u^{(j)} - \bar{\Delta}, \end{aligned}$$

where the vector $\bar{\Delta}$ is selected stochastically at any single energy splitting of two molecules according to the uniform distribution on

$$(11.2.3) \qquad \mathcal{D} := \left\{ \Delta \in \mathbb{U}^{(i)} \mid \sum_{k \in \{i,j\}} H_0(\bar{u}^{(k)}) = \sum_{k \in \{i,j\}} H_0(u^{(k)}) \right\}.$$

To distinguish this concept of a dynamics, investigated in Moeschlin, Grycko (2005b), from other concepts, we refer to the discrete dynamics SDMED (and *stochastically determined momentum exchange directions*).

Another selection rule also investigated in Moeschlin, Grycko (2005b) determines $\bar{\Delta}$ as the solution of the optimization problem

$$(11.2.4) \qquad \left| \langle \frac{\bar{\Delta}}{|\bar{\Delta}|}, e \rangle \right| = \max_{\Delta \in \mathcal{D}} \left| \langle \frac{\Delta}{|\Delta|}, e \rangle \right|.$$

In this case we refer to the discrete dynamics DDMED (with *deterministically determined momentum exchange directions*).

With the selection rule (11.2.4) for $\bar{\Delta}$, the classical momentum exchange vector e, which is possibly not an element of the lattice $\mathbb{U}^{(i)} = \mathbb{U}^{(j)}$, is approximated as closely as possible; condition (11.2.4) ensures the best approximation of the classical momentum exchange direction with respect to the lattice $\mathbb{U}^{(i)} = \mathbb{U}^{(j)}$.

For both discrete dynamics SDMED and DDMED the choice of $\bar{\Delta}$ from the set $\mathcal{D}$ guarantees the conservation of energy. Moreover, the total momentum is conserved and the momenta $\bar{u}^{(i)}$ and $\bar{u}^{(j)}$ after the energy splitting remain within the lattice $\mathbb{U}^{(i)} = \mathbb{U}^{(j)}$.

A reflection of a molecule i off the boundary of the container is defined causally as in Section 1.8; since the container B is rectangular, such a reflection keeps the molecule i within the space $\mathbb{U}^{(i)}$ of momenta.

11.3 Discrete Entropic Distributions

Let $\Delta u > 0$ be a constant carrying the unit $[kg \frac{m}{s}]$ of momentum and let

$$\mathbb{U}^{(j)} = \{\Delta u \cdot (\xi_1, \xi_2) \,|\, \xi_1, \xi_2 \in \mathbb{Z}\}$$

be the momentum space of the j-th molecule, $j = 1, \ldots, N$. The lattice

$$\mathbb{U}^N := \underset{j=1}{\overset{N}{\times}} \mathbb{U}^{(j)}$$

as the Cartesian product of the $\mathbb{U}^{(j)}$, represents the momentum space of the system of N molecules. According to Section 11.1, the Hamiltonian $H \colon \mathbb{U}^N \longrightarrow \mathbb{R}_+$ of the system is defined in an additively representable way by

$$(11.3.1) \qquad H(u) := \sum_{j=1}^{N} H_0(u^{(j)}), \qquad u = (u^{(1)}, \ldots, u^{(N)}) \in \mathbb{U}^N,$$

where the function H_0 is given by

$$(11.3.2) \qquad \begin{aligned} &H_0 \colon \mathbb{U}^{(j)} \longrightarrow \mathbb{R}_+ \\ &H_0(w) = v_0 \cdot (\kappa_1 \cdot |w_1| + \kappa_2 \cdot |w_2|), \quad w = (w_1, w_2) \in \mathbb{U}^{(j)}, \end{aligned}$$

with a fixed $v_0 > 0$ carrying the unit $[m/s]$ of velocity.

We now adapt Theorem 2.1.3 in order to determine the density of the entropy optimizer. Since the momentum spaces $\mathbb{U}^N$ and $\mathbb{U}^{(j)}$, $j = 1, \ldots, N$, are lattices, a natural choice for the dominating measure μ in Theorem 2.1.3 are the counting measures on $\mathbb{U}^N$ and $\mathbb{U}^{(j)}$ being σ-finite in the present context.

Let ζ_0 and ζ_N denote the counting measure on $\mathbb{U}^{(j)}$ and $\mathbb{U}^N$, respectively. Obviously

$$Z_0(\beta) := \int_{\mathbb{U}^{(1)}} \exp(-H_0(w)) \, \mathrm{d}\zeta_0(w) := \sum_{w \in \mathbb{U}^{(1)}} \exp(-\beta H_0(w)) < \infty$$

and

$$Z_N(\beta) := \int_{\mathbb{U}^N} \exp(-\beta H(u)) \, \mathrm{d}\zeta_N(u) = Z_0(\beta)^N < \infty$$

for $\beta > 0$. In view of the additive form of H_0, it follows that

$$
\begin{aligned}
Z_0(\beta) &= \sum_{(\xi_1, \xi_2) \in \mathbb{Z}^2} \exp\big(- \beta v_0 \Delta u \cdot (\kappa_1 \cdot |\xi_1| + \kappa_2 \cdot |\xi_2|) \big) \\
&= \sum_{\xi_1 \in \mathbb{Z}} \exp\big(- \beta \cdot v_o \cdot \Delta u \cdot \kappa_1 \cdot |\xi_1| \big) \cdot \sum_{\xi_2 \in \mathbb{Z}} \exp(-\beta \cdot v_0 \cdot \Delta u \cdot \kappa_2 \cdot |\xi_2|) \\
&=: Z_{0,1}(\beta) \cdot Z_{0,2}(\beta)
\end{aligned}
$$

for $\beta > 0$.

The density

(11.3.3) $$f_\beta : \mathbb{U} \longrightarrow \mathbb{R}_+$$

with

$$
\begin{aligned}
f_\beta(u) &= \frac{1}{Z_N(\beta)} \exp\big(- \beta H(u) \big) \\
&= \prod_{j=1}^N \frac{1}{Z_0(\beta)} \exp\big(- \beta H_0(u^{(j)}) \big) \\
&= \prod_{j=1}^N \Big(\frac{1}{Z_{0,1}(\beta)} \exp(-\beta v_0 \kappa_1 |u_1^{(j)}|) \Big) \cdot \Big(\frac{1}{Z_{0,2}(\beta)} \exp(-\beta v_0 \kappa_2 |u_2^{(j)}|) \Big)
\end{aligned}
$$

determines a probability measure P_β on $\mathbb{U}^N$, being the N-th power of a probability measure P_β^0 on $\mathbb{U}^{(1)} = \ldots = \mathbb{U}^{(N)}$: $P_\beta = f_\beta \zeta_N = (P_\beta^0)^N$. Hence a ζ_0-density f_β^0 of P_β^0 is given by

(11.3.4) $$f_\beta^0(w_1, w_2) = \frac{1}{Z_0(\beta)} \exp(-\beta v_0 \kappa_1 |w_1| - \beta v_0 \kappa_2 |w_2|).$$

The family $(P_\beta)_{\beta > 0}$ is the family of discrete entropic momentum distributions of the whole system of N molecules parametrized by β.

11.4 Temperature Estimates

Recall the Definition (11.1.1) of momentum space $\mathbb{U}^{(j)}$ of a particular molecule

$$\mathbb{U}^{(j)} = \left\{ \Delta u \cdot (\xi_1, \xi_2) \,|\, \xi_1, \xi_2 \in \mathbb{Z} \right\}.$$

The lattice $\mathbb{U}^{(j)}$ clearly has the Cartesian product structure

$$\mathbb{U}^{(j)} = \mathbb{U}_1^{(j)} \times \mathbb{U}_2^{(j)}$$

with

$$\mathbb{U}_i^{(j)} := \left\{ \Delta u \cdot \xi \,|\, \xi \in \mathbb{Z} \right\}, \quad i = 1, 2.$$

The counting measure ζ_0 on $\mathbb{U}^{(j)}$ is the product measure

$$\zeta_0 = \zeta_{0,1} \otimes \zeta_{0,2}$$

of the counting measures $\zeta_{0,i}$ on $\mathbb{U}_i^{(j)}$, $i = 1, 2$. According to Section 11.3, the entropic density f_β of the momentum distribution with the parameter β has the product structure

$$
\begin{aligned}
f_\beta(u) = &\prod_{j=1}^{N} \left(\frac{1}{Z_{0,1}(\beta)} \exp(-\beta v_0 \kappa_1 |u_1^{(j)}|) \right) \\
&\cdot \prod_{j=1}^{N} \left(\frac{1}{Z_{0,2}(\beta)} \exp\left(-\beta v_0 \kappa_2 |u_2^{(j)}| \right) \right)
\end{aligned}
$$
(11.4.1)

for $u = \left(u^{(1)}, \ldots, u^{(N)} \right) \in \mathbb{U}^N = \underset{j=1}{\overset{N}{\times}} \mathbb{U}^{(j)}$. The factors

$$(11.4.2) \qquad \frac{1}{Z_{0,1}(\beta)} \exp(-\beta v_0 \kappa_1 |u_1^{(j)}|) \quad \text{and} \quad \frac{1}{Z_{0,2}(\beta)} \exp(-\beta v_0 \kappa_2 |u_2^{(j)}|)$$

in the representation (11.4.1) of f_β can be interpreted as densities of the first and second components of the momentum vector $u^{(j)}$ w.r.t. the counting measures $\zeta_{0,1}$ and $\zeta_{0,2}$, respectively.

The factors (11.4.2) indicate that the distribution of the first and second component of the momentum vector $u^{(j)}$ are double geometric ones.

The expected value of the kinetic energy w.r.t. i-th momentum component is given by

$$
\begin{aligned}
\varepsilon_{0,i}(\beta) &:= \frac{1}{Z_{0,i}(\beta)} \sum_{\xi \in \mathbb{Z}} v_0 \cdot \Delta u \cdot |\xi| \cdot \exp\left(-\beta v_0 \kappa_i \Delta u |\xi| \right) \\
&= \frac{2 q_i(\beta) \cdot v_0 \cdot \kappa_i \cdot \Delta u}{1 - q_i(\beta)^2}
\end{aligned}
$$
(11.4.3)

where $q_i(\beta)$ is defined by

$$q_i(\beta) := \exp\left(-\beta v_0 \kappa_i \Delta u\right), \quad i = 1, 2.$$

Supposing the validity of the Entropy Principle, statement (a), two estimates of the accurate parameter β^* of the entropic distribution of momenta will be established.

By Formula (11.4.1) it is clear that the strong law of large numbers is valid for components of momenta: the sequence of averages

$$(11.4.4) \qquad \left(\frac{1}{N}\sum_{j=1}^{N} v_0\kappa_i|u_i^{(j)}|\right)_{N\in\mathbb{N}}$$

of the energies of the molecules w.r.t. the i-th component of momentum, which can be easily estimated based on empirical momentum data from a computer experiment, approximates the expected value

$$(11.4.5) \qquad \varepsilon_{0,i}(\beta^*) = \frac{2q_i(\beta^*) \cdot v_0 \cdot \kappa_i \cdot \Delta u}{1 - q_i(\beta^*)^2},$$

based on the entropic distribution if N is large, $i = 1, 2$.

If the Entropy Principle holds true, a consistent estimate $\hat{\beta}_i$ of β^*, is given implicitly equating (11.4.3) and (11.4.4):

$$(11.4.6) \qquad \frac{1}{N}\sum_{j=1}^{N} v_0\kappa_i|u_i^{(j)}| = \frac{2q_i(\hat{\beta}_i)v_0 \cdot \kappa_i \cdot \Delta u}{1 - q_i(\hat{\beta}_i)^2}$$

because the solution $\hat{\beta}_i$ depends continuously on the average (11.4.3). The corresponding estimates $\hat{T}_1, \hat{T}_2$ of the temperature $T^* := 1/(k_B \cdot \beta^*)$ are given by

$$\hat{T}_i := \frac{1}{k_B\hat{\beta}_i}, \quad i = 1, 2.$$

11.5 The Course of Experimentation

As a first step, in experiment E 11.1 we examine computer-experimentally the existence of an ergodic momentum distribution of a system of molecules subject to discrete dynamics SDMED, i.e. we examine whether the empiric momentum distribution stabilizes.

The projections of the empiric momentum distribution onto the horizontal and vertical axes are displayed. Also displayed for comparison are the corresponding accurate entropic distributions, i.e. the double geometric distributions, their parameters q being estimated based on the momentum data.

To examine in experiment E 11.2 the validity of the Entropy Principle for the discrete dynamics SDMED, the temperatures based on the momentum data projected onto the first (horizontal) and second (vertical) axis of the momentum space, respectively, are estimated.

Note that the statistical coincidence of these two temperatures is a necessary condition for the validity of the Entropy Principle.

In experiments E 11.3 and E 11.4 again the invariance of pressure and the equipartition of energy with respect to the micro-constituents are treated, respectively. There, our basis is the same dynamics as in experiments E 11.1 and E 11.2, i.e. the discrete dynamics SDMED.

In experiments E 11.5 and E 11.6 we repeat the experiments E 11.1 and E 11.2, respectively , where the discrete dynamics SDMED is replaced by the discrete dynamics DDMED according to the selection rule (11.2.3).

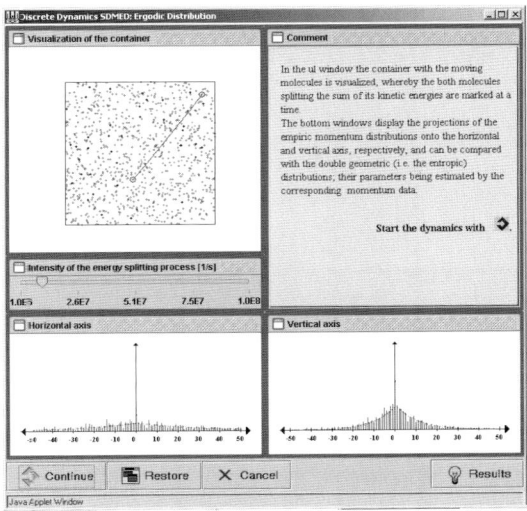

E 11.1. Ergodic distribution; discrete dynamics SDMED

11.6 Conclusions

The projections of the empiric momentum distribution onto the horizontal and vertical axis, respectively, become stabilized in the course of time for the dynamics SDMED; i.e. they are ergodic. Moreover they coincide statistically with the corresponding accurate entropic distributions. This can be recognized visually, but it is also confirmed by a χ^2-goodness-of-fit test at a level of significance of 0.05 performed by us. This means that the Entropy Principle is fulfilled for the case of the discrete dynamics SDMED.

This result is also confirmed by experiment E 11.2. The estimated temperatures based on the momentum data projected onto the horizontal and vertical axis, respectively, coincide statistically.

We already mentioned that, by the definition of the Hamiltonian H in Chapter 11, the double geometric distribution as an entropic distribution is related to the theoretical equilibrium probabilities of the excited energy levels of the harmonic oscillator from quantum mechanics, cf. Reif (1965) or Reif (1985).

The model of the discrete dynamics SDMED supports the theoretically postulated equilibrium probabilities of the harmonic oscillator.

Experiment E 11.3 does not confirm the invariance of pressure.

The equipartition of energy with respect to the micro-constituents is confirmed by experiment E 11.4.

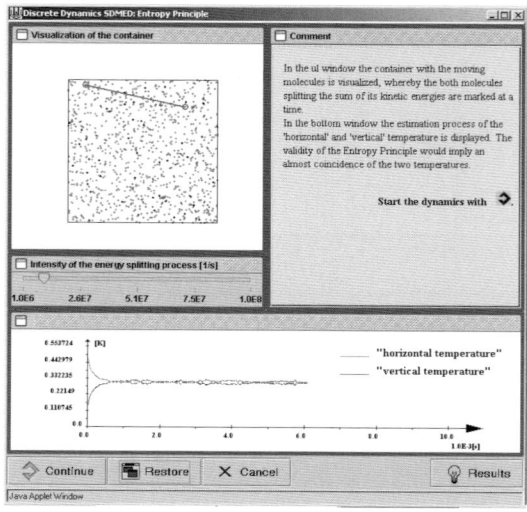

E 11.2. Entropy Principle (temperature); discrete dynamics SDMED

The discrete dynamics DDMED, with $\bar{\Delta}$ being selected according to (11.2.4), does not confirm the Entropy Principle; this follows visually from experiment E 11.5, but it is also confirmed by a χ^2-goodness-of-fit test, which we performed at a level of significance of 0.05.

Experiment E 11.6 yields two different estimates for the 'horizontal' and 'vertical' temperature, which contradicts the Entropy Principle.

Summarizing the results of Chapter 11, we have found that the discrete dynamics SDMED supports the theoretically postulated equilibrium probabilities of the harmonic oscillator of quantum mechanics. This discrete dynamics is a stochastical concept, the momentum exchange directions are at

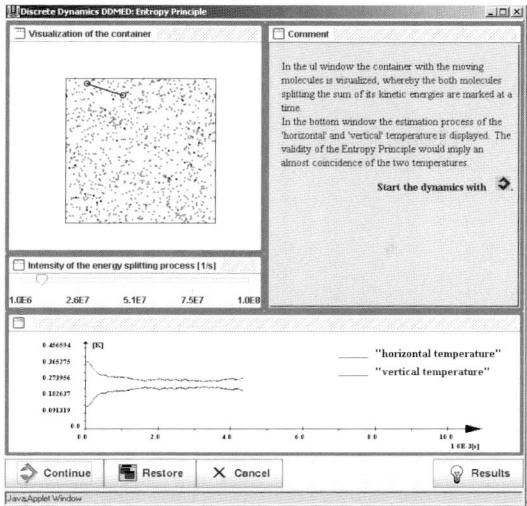

E 11.6. Entropy Principle (temperature); discrete dynamics DDMED

every single energy splitting outcomes of a random experiment according to the uniform distribution on set $\mathcal{D}$.

Moreover Chapter 11 shows that a deterministic concept to determine the momentum exchange direction – well-known from the Boltzmann system of moving molecules, as for instance the discrete dynamics DDMED – cannot explain the equilibrium probabilities of the energy levels of the harmonic oscillator.

Note, apart from pressure phenomena, where the hits of molecules off the boundary are of interest, that the configuration space, i. e. the container within which the molecules move, becomes obsolete. We refer to the final remark of Chapter 9.

12 Stochastic Energy Splitting

12.1 The Relevance of Momentum Conservation

In various experiments – beginning with the Boltzmann system of moving molecules – it was examined whether a specific definition of the energy or momentum function calls the Entropy Principle in question.

Indeed, momentum exchange according to the law of momentum conservation was in all the experiments an inherent principle determining the dynamics and in consequence the empiric momentum distribution. In all the investigated cases the Entropy Principle could be confirmed. So the question arises whether the validity of the Entropy Principle requires a dynamics based on momentum conservation. Note that the family of entropic distributions is defined by the Hamiltonian (as function of momentum).

The question to be treated in experiments E 12.1 and E 12.2 is whether the law of momentum conservation can be substituted by a stochastic rule, when splitting the energy.

12.2 A Stochastic Dynamics

In order to study the question of Section 12.1 for the Hamiltonian of the standard Newtonian dynamics, the energy splitting of two colliding molecules is organized in such a way that the kinetic energy of the one molecule is given by

$2t\,X$ (sum of the kinetic energies

of the colliding molecules before the impact) ,

while the kinetic energy of the other molecule is given by

$(1 - 2t\,X)$ (sum of the kinetic energies

of the colliding molecules before the impact) ,

where X is a random variable, uniformly distributed on $[0, 1]$ and $t \in \{0.02, 0.05, 0.1, 0.2, 0.5\}$ to be chosen by the experimenter. Note that for $t = 0.5$, the expected kinetic energies of the two molecules are identical.

According to the described model of energy splitting, the norms of the velocities of the two colliding molecules after an impact are calculated based on a random experiment determining a realization of the uniformly distributed random variable X.

Since velocity is defined as a vector, the polar angles of the velocity vectors of the two molecules still have to be fixed, which is also realized stochastically. To this end the experimenter has to select a distribution of the polar angles in degrees on $[0, 360]$.

12.3 The Course of Experimentation

In all the experiments of Chapter 12 the experimenter has first to choose the factor $t \in \{0.02, 0.05, 0.1, 0.2, 0.5\}$ in the ur window.

Then he either clicks on the mr window with the uniform distribution (Distribution I) of the polar angles of the velocity vectors, or on the mr window with the 'polygonal' distribution (Distribution II) of the polar angles. In the latter case the upper vertex of the 'polygonal' distribution can be shifted with the slider.

Note that in the window (Distribution II) and also in the ul window (Visualization), the median of the distribution of the polar angles is marked with a blue arrow.

When experimenting we suggest that the four cases should be distinguished

I   : $t = 0.5$, uniform distribution
II : $t = 0.5$, 'polygonal' distribution
III : $t \neq 0.5$, uniform distribution
IV : $t \neq 0.5$, 'polygonal' distribution

Experiments E 12.1 and E 12.2 follow the line of experiments E 7.1 and E 7.2, respectively. The aim of experiment E 12.1 is to prove the existence of an ergodic distribution and to compare the estimates with a centered normal distribution, i. e. the entropic distribution, its variance being estimated from the velocity data of the experiment.

Experiment E 12.2 calculates 'mechanically' the temperature as a function of the polar angle φ of the subspace L_φ. 'Mechanically' means that we do not first check whether the defining pre-conditions for the temperature are fulfilled.

Experiment E 12.3 also uses 'mechanically' the standard estimator for the temperature (being based on average kinetic energies) without checking if we really can speak of the notion of temperature.

12.4 Conclusions

Experiment E 12.1 shows that, apart from case I, a statistical coincidence of the empiric (ergodic) distribution and the accurate entropic distribution is not given; i.e. in the cases II, III, IV the Entropy Principle fails, with the consequence that a temperature as a scalar quantity cannot be defined.

For case I the coincidence of the empiric (ergodic) distribution and the accurate entropic distribution was tested applying a χ^2-goodness-of-fit test. At a level of significance of 0.05 the hypothesis that the ergodic distribution coincides statistically with the accurate entropic distribution, was not rejected.

This result of experiment E 12.1 is confirmed by experiment E 12.2; the latter follows the line of experiment E 7.2 to which we refer.

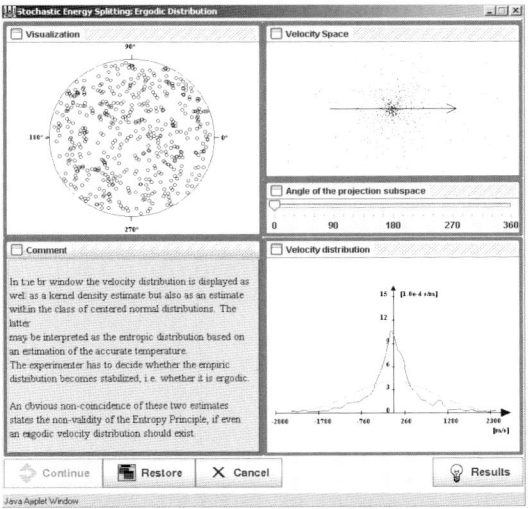

E 12.1. Ergodic distribution

Independently of the chosen value t of the factor of energy splitting, the invariance of pressure is given as long as the distribution of the polar angles of the velocity vectors is the uniform distribution, cf. experiment E 12.3. But note that, for $t < 0.5$, the Entropy Principle is not fulfilled, and therefore a temperature is not defined; if the prognosis by the equation of state coincides with the estimated pressure, this is because our estimation of temperature is based on the average energy.

The invariance of pressure does not hold true if the distribution of the polar angle of the velocity vectors is different from the uniform distribution.

The equipartition principle for energy is not fulfilled for $t < 0.5$, cf. experiment E 12.4; the isotonicity of the graph as generated in the bottom

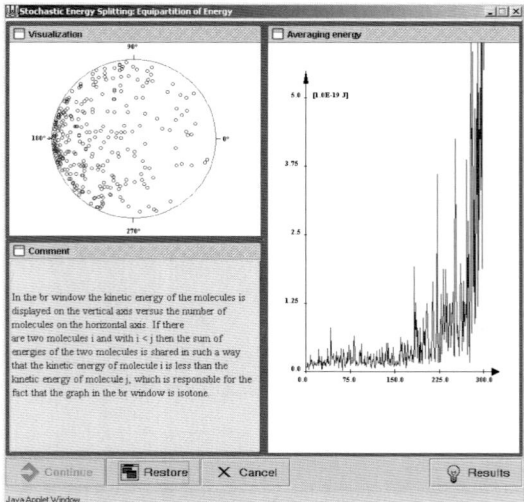

E 12.2. (Non-) Equipartition of energy

right window is of course a consequence of a programming detail. If there are two molecules i and j, $i < j$, then the sum of energies of the two molecules is shared in such a way that the kinetic energy of molecule i is less than the kinetic energy of molecule j; if we substitute this programming detail by a stochastic rule, the graph is no longer isotone but it is certainly not constant.

For $t = 0.5$ the equipartition principle of energy is fulfilled for any distribution of the polar angles of the velocity vector.

Bibliographical Comments

In Chapter 1 the Maxwell Hypothesis for velocities and momenta, which specifies the velocity and momentum distribution of a system of moving molecules subject to Newtonian dynamics, is presented. There the basic notions and facts from probability theory are presupposed as known, cf. for instance Bauer (1996).

The stochastic model of a fluid originating from the combined work of J.C. Maxwell and L. Boltzmann and also J.W. Gibbs is presented in most monographs on Statistical Mechanics. From the more recent literature we mention above others Gopal (1974), Greiner et al. (1995) and Schlögl (1989).

An intrinsic physical insight is the perception of the Maxwell Hypothesis; the variance of the so-called Maxwell–Boltzmann distribution is essentially the temperature. The merit of Ludwig Boltzmann to have recognized the relation between temperature and kinetic energy of a micro-constituent prompted us to bridge the conceptual discrepancies between the expected and the average energy of a micro-constituent, i.e. between the variance of the velocity distribution and its estimator arguing with the strong law of large numbers , cf. Bauer (1996), Theorem 12.1. The latter is important for our experimental-stochastical approach to a thermodynamical interpretation of computer-experimental data, cf. Moeschlin et al. (2003).

The dynamics of the moving molecules is based on the axioms of the Newtonian mechanics, cf. for instance Allonso, Finn (1979).

Comprehensive representations of statistical methods are given in Lehmann (1983), Moeschlin, Eberl (1982) or Zacks (1971). For kernel density estimators we refer to Devroye (1987) or Wertz (1978). A description of the χ^2-goodness-of-fit test may be found in Moeschlin et al. (2003). The statistics of linear models is treated for instance in Fahrmeier, Hamerle (1984).

Chapter 2 first presents the notion of the Boltzmann–Gibbs entropy of a probability measure, together with a theorem yielding a density of the optimizer of the Boltzmann–Gibbs entropy under the restriction that the expectation of a given non-negative, measurable function H is fixed, cf. Mackey (1992). The theorem mentioned may indeed be reformulated for other dominating measures than the Borel–Lebesgue one.

In various monographs of physics, cf. for instance Greiner et al. (1995), Reif (1965) or in the German translation Reif (1985), the density of the momentum distribution is given as an exponential term with the Hamiltonian in the exponent; a fact which – according to our knowledge – goes back to Max Planck. Theorem 2.1.3 about the entropy optimizer emphasizes the relation to the Boltzmann–Gibbs entropy.

Generalizing the Maxwell Hypothesis, we formulate in Section 2.3 an Entropy Principle which is motivated by the theorem characterizing the optimizer of the Boltzmann–Gibbs entropy.

In Grycko (2003) this concept was successfully applied to treat a system of micro-constituents being imposed on the relativistic dynamics and is used here to systematically study various dynamics, cf. Chapters 7–12.

The method of computer-experimentation is well-suited to investigate statistical aspects of a system of moving molecules. The subject of Chapters 3 and 4 is collision processes of molecules which, by a stochastical argumentation based on the axiomatic definition of the Poisson process, may be presumed as Poissonian, cf. Bauer (1996) or Doob (1953).

The computer-experimental analysis was part of a research project of the Department of Probability Theory and Statistics at the University of Hagen.

Without a precident is the investigation of the distribution of the reflection angles, cf. Section 3.5.

Pressure is explained as the time average of the molecular momentum absorbed by the boundary of the container. In Chapter 5 an estimator of pressure is investigated and practically applied to the analysis of the osmotic pressure for a dilute solution; (van't Hoff's law).These concepts were developed together with students of the Department of Probability Theory and Statistics.

Computer experimentation even allows us to analyze phenomena from non-real physics, i.e. artificially generated phenomena which have no real counterparts. Such an example is – with regard to later chapters – presented in Chapter 6. The example is taken from Grycko, Moeschlin (2000).

(M, M)-dynamics of Chapter 7 is a natural generalization of Newtonian dynamics, well-suited for the examination of the Entropy Principle, cf. Moeschlin, Grycko (2005a), which is done in Chapters 7–9 together with various alterations of the 'organizational forms' of momentum exchange.

Chapter 10 examines the validity of the Entropy Principle for a fluid being imposed on relativistic dynamics, cf. Grycko (2003). A general introduction to relativity theory can be found in Orear (1979) or in the German translation Orear (1982).

In Chapters 11 and 12 examples of dynamics are given for which the Entropy Principle does not hold true; the discrete dynamics of Chapter 11 is related to the often mentioned equilibrium probabilities of the excited energy levels of the harmonic oscillator from quantum mechanics, cf. for instance Reif (1965) or the German translation Reif (1985). In these chapters we report on computer-experimental experiences from the research project, Moeschlin, Grycko (2005b).

References

Allonso, M., E.J. Finn: Fundamental University Physics I, Mechanics. 2nd. Edition, Reading 1979

Bauer, H.: Probability Theory. Berlin, New York 1996

Cercignani, C., R. Illner, M. Pulvirenti: The Mathematical Theory of Dilute Gases. New York, Berlin, Heidelberg 1994

Devroye, L.: A Course in Density Estimation. Boston, Basel, Stuttgart 1987

Doob, J.L.: Stochastic Processes. New York, London, Sydney 1953

Fahrmeir, L., A. Hamerle: Multivariate statistische Verfahren. Berlin, New York 1984

Gopal, E.S.R: Statistical Mechanics and Properties of Matter. New York 1974

Greiner, W., L. Neise, H. Stöcker: Thermodynamics and Statistical Mechanics. New York, Berlin, Heidelberg 1995

Grycko, E.: Relativistic dynamics and the entropy maximization principle. Proceedings of the Fourth International IMACS Symposium on Mathematical Modelling in Vienna, 2003

Grycko, E., O. Moeschlin: An investigation of the Maxwell Hypothesis through methods from Experimental Stochastics. Transactions of Academy of Sciences of Azerbaijan, Series of Physical-Technical and Mathematical Sciences, Vol XXI, No. 1, pp. 222–231, 2001

Grycko, E., O. Moeschlin: A Generalized Newtonian Dynamics and the Maxwell Hypothesis. Seminarberichte FB Mathematik, Universität Hagen, Hagen 2005a

Grycko, E., O. Moeschlin: Entropy Principle and Momentum Exchange on a Discrete Momentum Space. Preprint 2005b

Jäckle, J.: Einführung in die Transporttheorie. Braunschweig 1978

Lehmann, E.L.: Theory of Point Estimation. New York, Chichester, Brisbane 1983

Mackey, M.C.: Time's Arrow: The Origins of Thermodynamic Behavior. New York, Berlin, Heidelberg 1992

Moeschlin, O., W. Eberl: Mathematische Statistik. Berlin 1982

Moeschlin, O., E. Grycko: An investigation of the Maxwell Hypothesis through methods from Experimental Stochastics. Transactions of Academy of Sciences of Azerbaijan, Series of Physical-Technical and Mathematical Sciences, Vol XXI, No. 1, pp. 222–231, 2001

Moeschlin, O., E. Grycko: A Generalized Newtonian Dynamics and the Maxwell Hypothesis. Seminarberichte FB Mathematik, Universität Hagen, Hagen 2005a

Moeschlin O., E. Grycko: Entropy Principle and Momentum Exchange on a Discrete Momentum Space. Preprint 2005b

Moeschlin O., E. Grycko, C. Pohl, F. Steinert: Experimental Stochastics. 2nd ed., Berlin, Heidelberg, New York 2003

Orear, J.: Physics. New York 1979

Orear, J.: Physik. München, Wien 1982

Reif, F.: Fundamentals of Statistical and Thermal Physics. New York 1965

Reif, F.: Statistische Physik und Theorie der Wärme. Berlin, New York 1985

Schlögl, F.: Probability and Heat. Braunschweig, Wiesbaden 1989

Wertz, W.: Statistical Density Estimation. A Survey. Göttingen 1978

Zacks, S.: The Theory of Statistical Inference. New York 1971

Symbols

General

$\Longrightarrow$	implication
$:=$	defining equality
$x \in A$	x is an element of the set A
$\square$	q.e.d., end of a proof
$x \notin A$	x is not an element of the set A
$E(x), \quad x \in A,$	$E(x)$ holds for all $x \in A$
$\{x|E(x)\}$	the set of all x satisfying $E(x)$
$\emptyset$	the empty set
$A \subset B$	A is subset of B
A^c	the complement of the set A
$\bigcup\limits_{i=1}^{n} A_i = A_1 \cup \ldots \cup A_n$	the union of sets $A_1, \ldots, A_n$
$\bigcap\limits_{i=1}^{n} A_i = A_1 \cap \ldots \cap A_n$	the intersection of sets $A_1, \ldots, A_n$
$\bigtimes\limits_{i=1}^{n} A_i = A_1 \times \ldots \times A_n$	the Cartesian product of sets $A_i, \ i = 1, \ldots, n$
$\partial(A)$	topological boundary of the set A
$\mathbb{N}$	the set of positive integers (natural numbers)
$\mathbb{Z}$	the set of integers
$\mathbb{R}_+ := \{x \in \mathbb{R}|x \geq 0\}$	the set of non-negative real numbers
$[a;b] := \{x \in \mathbb{R}|a \leq x \leq b\}$	for $a, b \in \mathbb{R} \cup \{-\infty, +\infty\}$
$[a;b) := \{x \in \mathbb{R}|a \leq x < b\}$	for $a, b \in \mathbb{R} \cup \{-\infty, +\infty\}$, etc.
$\mathbb{N}_n := \{x \in \mathbb{N}|x \leq n\}$	
$\mathcal{P}(A)$	power set of the set A, i.e. the set of all subsets of A

f^{-1}	the inverse mapping of mapping f, f being bijective
$f \circ g := f(g(.))$	composition of mappings f and g
1_A	indicator function of the set A
π_i	projection onto the i-th axis
$\langle ., . \rangle$	standard scalar product on $\mathbb{R}^n \times \mathbb{R}^n$
r.v.	random variable
r. ve.	random vector
$\mathcal{A}_1 \otimes \mathcal{A}_2$	the product of the σ-fields $\mathcal{A}_1$ and $\mathcal{A}_2$
$\mathcal{B}$	the Borel σ-field on $\mathbb{R}$
$\mathcal{B}^n$	the Borel σ-field on $\mathbb{R}^n$
$(\Omega, \mathcal{A})$	measurable space
$(\mathbb{R}^n, \mathcal{B}^n)$	measurable space with $\Omega := \mathbb{R}^n$ and $\mathcal{A} := \mathcal{B}^n$
$(\Omega, \mathcal{A}, P)$	probability space
P-a.e.	P almost everywhere
$f\mu$	the measure with μ-density f
$\bigotimes_{i=1}^{n} P_i := P_1 \otimes \ldots \otimes P_n$	the product of the probability measures $P_1, \ldots, P_n$
P_T	image of the probability measure P under the r.v. T
λ, λ^n	Lebesgue measure on $(\mathbb{R}, \mathcal{B})$ and on $(\mathbb{R}^n, \mathcal{B}^n)$
$\Pi(\lambda)$	Poisson process with parameter λ
π_λ	Poisson distribution with parameter λ
$\mathrm{Exp}(\gamma)$	exponential distribution with parameter γ
$N(a, \sigma^2)$	normal distribution with mean a and variance σ^2
$\mathbb{E}(X)$	expectation of r.v. X
$\mathrm{Cov}(X, Y)$	covariance of X and Y

Special

Subject Index